Mountain Farming Is Family Farming

A contribution from mountain areas to the International Year of Family Farming 2014

2013

Published by the Food and Agriculture Organization of the United Nations,
the Mountain Partnership Secretariat,
the Centre for Development and Environment of the University of Bern and
the Centre for Development Research of the University
of Natural Resources and Life Sciences, Vienna.

This publication was supported by the Austrian Development Cooperation,
the International Centre for Integrated Mountain Development,
the Swiss Agency for Development and Cooperation and
the World Bank through the Development Grant Facility

The designations employed and the presentation of material in this information product do not imply the expression of any opinion whatsoever on the part of the Food and Agriculture Organization of the United Nations (FAO) concerning the legal or development status of any country, territory, city or area or of its authorities, or concerning the delimitation of its frontiers or boundaries. The mention of specific companies or products of manufacturers, whether or not these have been patented, does not imply that these have been endorsed or recommended by FAO in preference to others of a similar nature that are not mentioned.

The views expressed in this information product are those of the author(s) and do not necessarily reflect the views or policies of FAO.

ISBN 978-92-5-107975-1 (print)
E-ISBN 978-92-5-107976-8 (PDF)

© FAO 2013

FAO encourages the use, reproduction and dissemination of material in this information product. Except where otherwise indicated, material may be copied, downloaded and printed for private study, research and teaching purposes, or for use in non-commercial products or services, provided that appropriate acknowledgement of FAO as the source and copyright holder is given and that FAO's endorsement of users' views, products or services is not implied in any way.

All requests for translation and adaptation rights, and for resale and other commercial use rights should be made via www.fao.org/contact-us/licencerequest or addressed to copyright@fao.org.

Editors: Susanne Wymann von Dach (CDE), Rosalaura Romeo (FAO/MPS), Alessia Vita (FAO/MPS), Maria Wurzinger (BOKU), Thomas Kohler (CDE)

Authors of case studies and introductory texts: international group of experts (for the names see list of authors)
Concept: FAO Mountain Partnership Secretariat, CDE, BOKU
Layout: Simone Kummer (CDE)
Editing: Nancy Hart, Sara Manuelli, Mia Rowan
Proofreading: Stefan Zach

Citation:
Wymann von Dach S, Romeo R, Vita A, Wurzinger M, Kohler T (eds). 2013. Mountain Farming Is Family Farm-ing: A contribution from mountain areas to the International Year of Family Farming 2014. Rome, Italy: FAO, MPS, CDE, BOKU, pp. 100

This publication is available from:
FAO information products are available on the FAO website (www.fao.org/publications)
and can be purchased through publications-sales@fao.org.

Electronic version can be downloaded from:
www.fao.org
and
www.mountainpartnership.org

Cover photo: Hmong ethnic hilltribe families harvesting rice on a terraced rice field in the northern mountainous province of Yen Bai, Viet Nam (Hoang Dinh Nam, AFP/Getty Images)

Contents

Foreword	5
1 Mountain farming is family farming	**10**
2 Global change and mountain livelihoods	**14**
Transformation of mountain livelihoods	16
Crisis offers chances for tourism and organic farming	18
Between melting glaciers, a growing metropolis and the world market	20
Farming on the fringe: adaptation to urbanization	22
3 Learning and cooperation	**26**
Building on traditional cooperation among women	28
A farmers' cooperative and a supermarket team up	30
Radio Mampita – the powerful voice of rural people	32
A school for promoters of agro-ecology	34
Field schools for agro-pastoralists	36
Lobbying for mountain regions and farming	38
4 Sustainable intensification and organic farming	**42**
Towards a fully organic state	44
Kitchen gardens for improved well-being	46
Organic farming improves income and diet	48
Sustainable mountain pastoralism: challenges and opportunities	50
Improvement of aquaculture practices in mountain farming	52
Organic farming as a climate change adaptation measure	54

5 Mountain products and market development — 58

- Certification frameworks for mountain products — 60
- Agribusiness development through cooperation — 62
- Adding value to traditional mountain crops — 64
- Spinning a fine yarn — 66
- Community-based beekeeping for better livelihoods — 68

6 Diversification of mountain livelihoods — 72

- Diversification – a historical perspective — 76
- Small forest-based enterprises reconcile conservation and development — 78
- Social agriculture as part of green care — 80
- Rural tourism promotion builds on local values — 82

7 The future of family farming in mountains: policy messages — 86

Authors and editors — 88

References and further reading — 92

Foreword

Mountain farming is largely family farming – which for centuries has contributed to sustainable development. Thanks to its small-scale character, diversification of crops, integration of forests and husbandry activities, and low carbon footprint, mountain agriculture has evolved over the centuries in an often harsh and difficult environment. The lifestyles and beliefs of mountain communities have inspired them to seek sustenance from the land but also to conserve the natural resource base and ecosystem services vital to downstream communities both rural and urban.

However, recent trends in global development have significantly reduced the resilience of mountain ecosystems. Increasing population, climate change, deforestation, desertification, market integration as well as changes in human values and aspirations are all taking a heavy toll on mountains and mountain development. Yet, in a world increasingly aware of "green" quality and organic products, mountain agriculture can provide high-value and high-quality products that cater to increasing market demand and generate income for local communities.

This growing attention to family farming presents an opportunity for mountain farmers to receive greater support and specific policy interventions. Family farming encompasses all the activities within the realms of agriculture, forestry, fisheries, pastoralism and aquaculture that are predominantly reliant on family labour. The General Assembly of the United Nations proclaimed 2014 as the International Year of Family Farming to recognize and support the contribution of family and smallholder farms to food security, poverty eradication and achieving the Millennium Development Goals.

To tap the potential of mountain agriculture fully, mountain communities would benefit from targeted support for strengthening the value chain – from planning and producing to processing and marketing. An enabling policy environment that encompasses tailored investments, business development and financial services is necessary to improve mountain farmers' access to resources and enhance their capacities to generate income. Support to mountain farming and the creation of new, diversified employment, training and educational opportunities should be embedded in all (sub)national mountain development policies. Unless the livelihoods of mountain communities improve, local people will continue to migrate to lowlands, cities or other countries. The loss of traditional landowners could leave mountain areas to those who will not have the same knowledge or commitment to use the land in a sustainable way, meaning increasing risk for key ecosystem services such as water and soil management, and biodiversity conservation, which could have add-on risks that would not only affect the mountain people but also the populations in the plains and cities.

This publication intends to raise awareness of the importance of mountain family farming in sustainable development worldwide and encourage investment in this sector. To mark the International Year of Family Farming 2014, the Food and Agriculture Organization of the UN, the Mountain Partnership Secretariat, the Austrian Development Cooperation, the International Centre for Integrated Mountain Development, the Swiss Agency for Development and Cooperation, the Centre for Development and Environment of the University of Bern and the Centre for Development Research of the University of Natural Resources and Life Sciences, Vienna have jointly issued this publication. Mountain Farming Is Family Farming is published at a time when the Post-2015 development agenda is being discussed. It is our aspiration that issues related to sustainable mountain development are adequately reflected in the UN Sustainable Development Goals and the Post-2015 development agenda. The following chapters, with concrete case studies, showcase the sustainable development of mountain communities and environments, a cause to which all of the co-publisher organizations are committed.

Eduardo Rojas-Briales

Assistant Director-General
Forestry Department – FAO

Mountain farming is family farming

Three generations are threshing barley in Pitumarca, Peru (S.-L. Mathez-Stiefel)

Family enjoying a rest during a hard working day in their small *chacra* (field), Bolivia (S.-L. Mathez-Stiefel)

Thomas Kohler and Rosalaura Romeo

Mountain farming is family farming

From a global perspective, mountain farming is family farming. Mountain areas, with their dispersed patches of useable land at different altitudes with different climates and with their often highly fragmented landscapes and narrow limits for mechanization, are most efficiently and effectively managed by family farms.

Family farming in mountains is as diverse as the myriad mountain landscapes of the world, but at the same time, there are also commonalities. For example, mountain family farms are usually not the centres of national production in terms of quantity, with the exception of tropical mountain regions. Most of their production is for family consumption, playing a key role in ensuring household food security. In addition, family farms in mountains help shape mountain landscapes, providing ecosystem services that are vital for development far beyond mountain areas. These services include provision of freshwater, disaster risk reduction, preservation of biodiversity including agro-biodiversity, and space for recreation and tourism.

Family farming communities also are custodians of place identity, spiritual and cultural values, and of site-specific knowledge – a precondition for survival in most mountain areas. The motivation of family farmers thus goes beyond profit maximization, to include social, cultural and ecological motives (1). This is particularly important in mountain areas, where time and resources required for reproductive activities – those that do not directly generate income but are indispensable for maintaining the natural production base – are generally higher than in lowland areas. The terraced landscapes found in all major mountain regions of the world are the most spectacular testimony of such reproductive investment. In addition, family farming in mountains largely operates with low external inputs, most often owing to circumstances rather than choice, meaning that mountain farmers often do not have the means, in terms of physical access or finance, to invest in

A family preparing their land for winter wheat, Tajikistan (B. Wolfgramm 2006)

external inputs such as fertilizer, plant and animal protection chemicals, let alone machinery.

Accessibility is a key issue in mountain farming, especially in developing countries. But this goes far beyond access to farm inputs – it includes access to basic infrastructures such as health services, schools, roads, transport, markets and communication with the outside world. This lack can be attributed to difficult topography and low population densities relative to lowland areas, factors that increase investment and maintenance costs. Moreover, mountain farmers – like mountain people in general – are often a minority in their countries in terms of numbers. They live far away from the centres of economic and political power and decision-making, and are often marginalized in political, social and economic terms. This is particularly true for communities with livelihoods and farming practices that deviate from global and national mainstreams, such as shifting cultivators or pastoralists, which are both prominent and important in mountain regions. Pastoralists, for example, use large tracts of marginal mountain lands through mobility that would remain unproductive otherwise.

One of the results of marginalization is widespread poverty. Around 40% of mountain populations in developing and transition countries – about 300 million people – are food insecure, with half of them suffering from chronic hunger (2). In response, family farming in many mountain areas is increasingly affected by outmigration. Although those who leave can provide remittances, it also means heavier workloads for those remaining – women, children and the elderly. Limited availability of land that often has low productivity, lack of recognized land tenure rights and population pressure are all elements that can contribute to unsustainable use of mountain natural resources.

The International Year of Family Farming (IYFF) 2014 presents an opportunity to focus attention on the merits and challenges of family farming in mountain areas. Supporting sustainable forms of family farming also promotes food security and a balanced diet and good environmental stewardship. This also recognizes and supports values and traditions that are conducive to securing key ecosystem services that are critical for development and that reach far beyond mountain regions. In mountain areas, family farming often remains an occupation of last resort while, under the right conditions, it could become the backbone for sustainable development. This report highlights examples from mountain areas worldwide that have made inroads towards this aim.

What is family farming?

According to FAO's working definition, family farming is a means of organizing agricultural, forestry, fisheries, pastoral and aquaculture production that is managed and operated by a family and predominantly reliant on family labour, including both women's and men's. The family and the farm are linked, co-evolve and combine economic, environmental, social and cultural functions.

Family farming is one of the most predominant forms of agriculture worldwide, in both developing and developed countries. Diversity of national and regional contexts, in terms of agro-ecological conditions, territorial characteristics, infrastructure availability (access to markets, roads, etc.), policy environment and demographic, economic, social and cultural conditions, influence family farming structures and functions, as well as livelihood strategies. [3]

Globally, the sector employs 2.6 billion people or 30% of the world's population, and is especially important in developing countries. While family farming covers a wide spectrum of farm sizes and types, ranging from large mechanized farms to smallholdings of a few hectares or less, it is the small family farms, run by small producers, that are by far the most numerous. Globally, they account for about 99% of all people engaged in farming [4].

www.fao.org/family-farming-2014/en/

Global change and mountain livelihoods

2

El Alto, a growing metropolis on the Altiplano of Bolivia (D. Hoffmann)

Farmers in Murghab District, Tajikistan, integrate new technologies with traditional living (B. Wolfgramm)

Hans Hurni

Global change and mountain livelihoods

Global change that affects mountain environments has many facets, extending beyond the biophysical impacts on temperatures, extreme weather events, melting glaciers and shortened snow cover related to climate change. Global change also has profound impacts on forest cover and composition, land use patterns and systems, water cycles and qualities, soil health and degradation, and agrobiodiversity. And even more broadly, many changes occurring globally have profound socio-economic impacts on mountain people.

The lives and livelihoods of mountain people are affected by the same socio-economic changes that affect people in the rest of the world, although impacts are often more profound, owing to the increased vulnerability and reduced resilience of mountain environments. These socio-economic changes that can affect mountain people both positively and negatively, include economic globalization, increasing accessibility, dynamic demography, more social infrastructure and changing consumption patterns (Figure 1).

In developing and transition countries, mountain people have reduced possibilities. Their lack of good roads increases transaction costs, the steeper slopes on the farmland add to the cost of maintaining agricultural systems, and there are higher production and reproductive costs. In addition, they are disadvantaged owing to low current investment in, and less innovation adapted to, mountain farming conditions. Mountain farmers also have to deal with the fact that 17% of mountain areas outside Antarctica are "protected areas", which has potentially negative effects on mountain farming due to banning or restricting farming activities.

Global change in mountains may also lead to disadvantages for livelihoods. For example, if men must migrate in search of labour opportunities outside the mountains, it may result in feminization of mountain farming. Unless sufficient labour can be mobilized for farm activities and for maintaining the stability of natural resource use, children may be taken out of school to work on family farms. Also, if there is an insufficient labour force, the terrace systems that enable farming in steep mountain areas can disintegrate in a very short time period.

Yet, mountain areas often have access to water for irrigation or drinking water supply. Equally, mountains may be favourable areas owing to their potential for tourism development, which is often coupled with conservation areas owing to higher biodiversity – an asset for tourism. And those relatives who have migrated can still support their families and their mountain communities through increased remittances.

Equally, with the global trend towards better access and social services, urbanization and market integration are now taking place in mountain environments. Although often at a slower pace than in lowlands, these trends contribute to improving the livelihoods of mountain communities and help integrate them into national and regional markets. Mountains are also used increasingly by urban populations for recreation and leisure, thereby offering mountain communities an opportunity to move from subsistence to cash crop and livestock production, and away from primary occupation to services.

Remittances flowing to Kathmandu, Nepal, drive urbanization in the most fertile land (S. Wymann)

Although global change has both positive and negative impacts, the issue is that the negative consequences may be more pronounced in mountains, both for the communities and for their environments, requiring more awareness, more attention and quicker reaction than elsewhere. Equally, the consequences of negative impacts may go beyond the boundaries of mountains and affect people and ecosystems in the surrounding lowlands. While water is the most obvious resource for explaining such interactions, there are many more concerns, such as unwanted migration, negative impacts of reduced snow and ice cover, a loss of quality of agricultural products from mountains, or reduced potential for tourism and recreation. International cooperation in sustainable mountain development and international cooperation in research, education and knowledge generation have the potential to help identify changes with negative implications for mountain livelihoods and resources. This, in turn, may lead to finding long-lasting solutions to such problems, while strengthening the ability to benefit from positive potentials for sustainable mountain development as they emerge from global change processes.

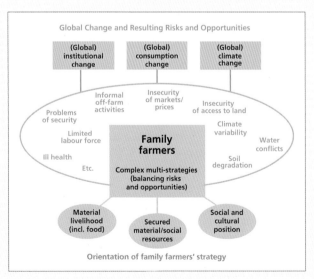

Figure 1: Family farmers aim to balance risks and opportunities that come with global change
Source: (1), modified

Transformation of mountain livelihoods

Bob Roga Nakileza and Peter Mukwaya

Mountain farming in East Africa has been profoundly transformed over the last decades, due to government policies, population growth, land scarcity and dwindling farm size, outmigration and an increasing integration into world commodity markets. This transformation is especially visible in the areas of Mount Elgon in Uganda and the Rungwe Mountains in Tanzania.

Farmers marketing their produce at Kamu in Mount Elgon, Uganda (B. Nakileza)

Mountains and highlands in East Africa have great potential as farming areas – unlike mountain areas in temperate zones. Rainfall is higher and more reliable than in the lowlands, and soils are generally fertile. Covering some 19% of the land area of Uganda and 23% of Tanzania and Kenya, mountains and highlands are home to the majority of the population and include major urban areas. Smallholder family farms in these areas are the most important food producers and thus critical for ensuring regional food security. Yet today, the farm sizes, generally below 1–2 ha, are decreasing even further owing to increasing population densities.

The drivers of change are many (Table 1). Agricultural policy geared towards modernization, the widespread use of mobile phones, radio and TV, the increased development of roads, the growth of small urban centres in rural areas and the attraction of youths towards urban lifestyles have deeply altered rural livelihoods. However, globalization is the main driver.

Traditionally, mountain farmers in East Africa produced for subsistence, but in late colonial times and especially after independence in the 1960s, they increasingly engaged in commodity production, growing crops such as barley, wheat, coffee and tea. Since the early 1990s, their horticultural products such as vegetables and flowers, which are largely sold on the European market, have brought good prices and provided quick cash, while diversifying farm production. Households now depend on the prices paid for these commodities on national and world markets, and on the institutions handling the products. In addition, changes in climate, such as reduced or less reliable rainfall, are reported across the area, but solid evidence of the impact is difficult to ascertain.

"*M*ost working class of today have grown up and been educated with farm proceeds. Unfortunately, many never return but stay in town or buy land elsewhere. This represents a loss to the family farm and to its human capital."

The Honourable Bernard Wolimbwa, local farmer and former Member of Parliament, Mount Elgon area, Uganda

The interplay of the above factors has transformed the traditional family farm profoundly.

- Fewer social assets and weakened social cohesion within an extended family and rural community – as a result, the mobilization of community resources has become much more difficult.

- Smaller sized farms constrain agricultural production and hence incomes – as a result, poverty in East Africa's mountains reaches 50–60% in spite of their high potential (1), and rural food insecurity is high.

- Less use of external inputs, such as fertilizers and pesticides, with the exception of commercial crops, due to non-enabling environments – as a result, in the Rungwe Mountains, for example, tractor hire services must be paid upfront, which makes them inaccessible to most small farmers. This affects household and regional food security negatively (2).

Many initiatives in the two regions address these issues, led or launched by different actors including the government, international development agencies, church groups and civil society institutions, including local groups. They mainly aim to enhance local financial and social capital. This can include village banks; farmers' and women's groups for knowledge exchange, credit and saving facilities, and marketing; cultural groups for safeguarding local traditional knowledge; and extension services of the government. Efforts are also made to further diversify farmers' production portfolio. Agroforestry is promoted as a measure to curb soil erosion and intensify production sustainably. Zero grazing is increasingly practised as a response to land scarcity and degradation, and to ease the collection of manure for improving soil fertility.

Lessons learned

- Land pressure due to increasing rural population density and high levels of poverty could undermine the sustainable use of these highly productive and high-potential mountain agro-ecosystems (3).
- Knowledge and youth: the out-migration of youth may help ease this pressure, but represents a loss of active human capital for rural areas. This may negatively affect the innovative potential of these areas, including family farms.

Fragmented land on slopes in the Rungwe Mountains, Tanzania (B. Nakileza)

Drivers of change	Effects in region and on family farming
Government policy and institutions	Collapse of villagization policy; adoption of market economy (Rungwe). Liberalization and collapse of the cash crop (coffee) economy; reduced capability of cooperatives to offer subsidies (Mount Elgon).
Demography and land	High and increasing population density; decrease and fragmentation of farmland.
Migration	Male and youth outmigration for education and jobs; devaluation of farm work as compared with work in industry or services; feminization of farming.
Infrastructural development	Increased density of (all-weather) roads.
Urbanization	Many small centres emerging, that create markets and provide services; rural–urban exchange is increasing, rural–urban gradient decreasing.
Information technology	Increased use of mobile phones and Internet (both regions); radio broadcasts reaching out to farmers (Mount Elgon).
Economic globalization	Cash crop production increases households' product portfolio, but also their vulnerability to price fluctuation on world markets.
Climate variability and change	Rainfall variability and periodicity changing; pests and diseases increasing.
Women are members of interest groups	37%

Table 1: Drivers of change in family farming: the examples of the Rungwe Mountains (Tanzania) and Mount Elgon (Uganda)

Crisis offers chances for tourism and organic farming

Jelena Krivcevic

Montenegro, a small, mountainous country on the Mediterranean Sea in southeast Europe, faced a severe economic downturn since the 1990s, due to political turnabouts, that impacted mountain farmers' livelihoods. However, today, Montenegro's increased tourism has opened new opportunities for enhancing organic farming in Montenegro's mountains.

Bjelasica mountain in Montenegro, tourism attraction and home to many farmers (J. Nikolic)

Montenegro's mountains have traditionally been home to farmers accustomed to difficult conditions and the need for hard work to survive the harsh winters. When industrialization set in after World War II, many farmers left their villages and moved to towns to work in factories. In the 1990s, most of these factories went bankrupt, leaving thousands without jobs. This major economic crisis, mainly triggered by the demise of former Yugoslavia, hit Montenegro hard and hampered any significant investments in mountainous areas. This led to dilapidated infrastructure, difficult supply of water and electricity, and poor access due to bad roads. Of course, all of this reduced interest of people in returning to their villages and engage in farming. Now, however, this situation is slowly changing.

It turns out that the crisis has had positive outcomes. For example, the closing of industries decreased pollution, and the reduced use of artificial fertilizers and farming chemicals also reduced soil contamination. Farmers in Montenegro use less than a tenth of the chemical inputs per hectare compared with farmers in the European Union (1). Along with the stunning beauty of the mountains, this has been recognized as a major advantage for tourism promotion and organic farming development.

In Montenegro, 6% of the population is engaged in agriculture, but the sector only contributes 0.8% to the gross domestic product, and 18% of people in rural areas are poor. At the same time, tourism contributes about 25% to the GDP – a share that is increasing (2,3). The Regional Development Agency (RDA) for Bjelasica, Komovi and Prokletije in northeastern Montenegro aims at improving mountain farmers' situations by taking advantage of the boom in tourism. Estab-

"*I would never believe that my products would attract attention of a buyer such as Porto Montenegro. This project has helped me in finding good buyers, designing very attractive packaging for my products, raising visibility for us small farmers in Montenegro, and I am very happy about this. The project has built our confidence, and the next thing I want to try is agrotourism; my wife and I want to set up bed & breakfast in our old house, and benefit from living in an attractive area with many hiking and biking trails around. We can offer our products, home-made dishes and authentic rural experience.*"

Milan Kljajic, farmer in Berane

Customers interested in regional products (J. Nikolic)

Lessons learned

- Major changes due to political transition or economic crisis can open up new opportunities if addressed in an appropriate and sustainable way.
- Without external support, it is difficult for small organic producers to fulfil the demanding requirements of certification schemes and to access new markets in cities or major tourist locations, which are often the first markets for organic products.
- Successful organic farmers will generate enough income, stay with farming and ensure steady rural development and, in turn, help reduce the significant development gap between mountains and other areas – such as coastal areas in the case of Montenegro.

lished in 2009, RDA has received funding from the Austrian Development Agency to explore opportunities for agrotourism, assist farmers in regional branding of organic and traditional products, and establish linkages with consumers in cities and tourist resorts.

Based on a local assessment, RDA selected 20 farmers who already produced organic and typical products and were interested in participating in a pilot project aimed at creating a regional brand. Until then, producers had been struggling to comply with all the rules required by Monte Organica, the certification institution in Montenegro. Compliance meant high costs for producers, while the market was still not ready to pay more for organic products. A chain of health food shops expressed interest in the project and arranged a space in one of its shops to present its organic mountain products as a speciality. RDA hired a design company to develop packaging and labelling ideas acceptable to farmers, but also a special design of the shop zone where the products would be located. Significant efforts were made in promotional activities. Overall, the project proved to be a major success. Soon after, shop owners in Porto Montenegro, a luxurious yachting marina, approached RDA to find ways to replicate this model within their resort. Given the popularity of Porto Montenegro and given the number of visitors they have each year, this can become a dream opportunity for mountain farmers.

A local honey bee producer (J. Nikolic)

Between melting glaciers, a growing metropolis and the world market

Nevado Huayna Potosí with the cemetery of Milluni mining village, Bolivia (D. Hoffmann)

Dirk Hoffmann, Liz Lavadenz, Rodrigo Tarquino

The Tuni Condoriri region of Bolivia's Cordillera Real is undergoing a fundamental change in an environment characterized by glacier recession, climate change, the growing nearby twin towns of El Alto and La Paz, and the proliferation of mining activities due to increased demand on the world market. All of this combines to impact the livelihoods of the region's family farms.

Bolivia's Tuni Condoriri region is home to indigenous Aymara communities that have practised livestock herding and subsistence agriculture in the area for centuries. Over the past two decades, as the region has been affected by global warming, urbanization and a boom in mining, so have the farmers' livelihoods.

The most visible evidence of global warming is the melting of the glaciers across the region, which covers 730 km² – of which about 2% (14.5 km²) are glaciers and 1% (7.4 km²) wetlands at an altitude of over 6,000 metres. In fact, Bolivia's Cordillera Real has lost about half of its glaciers over the last 35 years; a process that continues unabated, with most of the smaller glaciers forecast to disappear within the next 20 or 30 years.

At the same time, the region provides almost all the water for El Alto and about half of the water for La Paz, a metropolitan area that counts around 2 million people. About 12–15% of its drinking water is of glacier origin. Considering the ever increasing urban demand for freshwater, the loss of the glaciers could put an additional strain on the already scarce water supplies, exacerbating the potential for urban–rural conflicts over water rights and use.

Global warming has also provoked changes in rainfall patterns, with rains becoming so much less reliable that a number of farmers have stopped rainfed production. Moreover, the reduction of continuous periods of freezing has led families to stop producing the traditional *chuño*, a freeze-dried potato destined mainly for local consumption.

"*In the past it was colder, now there is a lot of sun, that is why the glaciers are melting rapidly, and there isn't as much water as before; but when it is raining, then there is a lot of water. Because of the heat we now have some mosquitoes that we didn't have before.*"

Small farmer from Tuni Condoriri region

"*In the past, there was more water and more wetlands, that is why the pastures were maintained throughout the dry season, but now, even the flow of our creeks has diminished, the rainy season is two months late, which means we have to postpone bringing out the seeds.*"

Small farmer from Tuni Condoriri region

The Millní cooperative mine (D. Hoffmann)

Lessons learned

- Family farms and communities in the Tuni Condoriri region have displayed a high capacity for dealing with the risks associated with farming, due to traditional knowledge systems and a high degree of internal organization and social cohesion that make them less vulnerable to the effects of global change.

- It remains to be seen whether current development trends, including the notion of what is a good life, will eventually lead to a loss of internal organization and cohesion, and whether communities will maintain their adaptive capacity to change in future.

As farming becomes less important, mining and linkages with the fast-growing twin towns of El Alto and La Paz offer important additional economic revenues for the majority of the region's rural population. As many younger people abandon the traditional mountain communities and migrate to the towns, it indicates the changing perspective of what constitutes *bien vivir*, a good life. As a result, El Alto has more than doubled its population over the last 20 years (Table 1).

Mining has become more prominent during the past decade. Increased prices for minerals on the world market have led to the re-opening of old mines and to the exploitation of new mines within the area. The sector is led by medium and small enterprises operated by local groups, mining mainly zinc and gold. The dynamic nature of mining is shown by the constantly high level of mining concessions granted. Since 2008, 75–100 new concessions have been granted every year, which means that every year between 15 and 18% of the region have been affected by new mining concessions. In total, an estimated one-third of the whole area is granted to mining enterprises; older concessions run out while new areas are opened up for exploitation (1).

A flock of alpacas, the preferred livestock of the region (D. Hoffmann)

Farmers' strategies to cope with global change still employ traditional patterns of risk management. Different altitudinal belts are still used for herding and for growing a diversity of crops, thus minimizing the risk of total failure. Employment in urban areas and in the mining sector has enabled families to follow a new strategy, with individual members of extended families pursuing different economic activities at different times of the year. This diversifies risks and opportunities, reducing dependence on local and non-predictable factors such as weather and climate, while increasing dependence on non-farm economies of regional and global scope.

City	1992	2001	2005	2010
La Paz	713,378	789,585	834,848	835,361
El Alto	405,492	647,350	795,740	953,253

Table 1: Population data for La Paz and El Alto. Data for 2005 and 2010 are projections based on the 2001 Census.
Source: National Statistics Institute (INE)

Farming on the fringe: adaptation to urbanization

Former grasslands in the *puna* altitudinal zone are increasingly converted into arable land for cultivating tubers at approximately 4,000 m asl in the Mantaro Valley, Peru (A. Haller)

Andreas Haller and Oliver Bender

The Central Andes are one of the world's most populated mountain regions. Their Quechua altitudinal belt, which runs between 2,300 and 3,500 m, was once dominated by farming hamlets, rural villages and small commercial towns. However, in recent decades, this region has undergone rapid urban growth that has profoundly altered land use and the livelihoods of peri-urban smallholders.

The growth of cities in Peru's Quechua belt (1) has become an important driver of land use change. For example, the population in Huancayo, located at 3,300 m, grew from 307,000 to 361,000 between 2000 and 2013 (2). Factoring in the surrounding peri-urban settlements brings the metropolitan area population to 425,000 (3).

Urban growth has its consequences for land use and livelihoods of smallholders who live on the rural–urban fringe. In Huancayo, the increasing demand for land and water has increased resource scarcity in the valley floor, the most favourable area for agricultural production, and has driven up land prices. Many smallholders of the Quechua belt own very small plots (5), which they mostly use for subsistence production. Thus, they depend on renting additional land for production of market-oriented crops, such as the maize, potatoes or artichokes that provide them with cash income. However, today's rising land prices have diminished smallholders' possibilities for renting such additional plots. In fact, the landowners, mostly large real estate firms, are not willing to let the land to farmers, fearing it might restrict their ability to develop and sell their property at the best moment in time.

Local smallholders perceive the urbanization of Huancayo's hinterland as a threat to their food and income security. Many of them cope by increasing production of home-based breeding of small animals such as guinea pigs, and selling the meat on urban markets (Table 1). They are also expanding or intensifying crop production on nearby community-owned slopes and high plains of the *suni* (3,500–4,000 m) and *puna* (4,000–4,800 m) altitudinal belts, trying to compensate for what they have lost on the valley floor. However, apart from potatoes, the crops grown in

"*Our nature is being more and more destroyed. Nowadays, concrete is sown and we will pay for this in future because the arable land is drastically reduced! Those who sell land for construction are not aware of the damage they cause and of where the food for our village will come from. Many residential projects are constructed by real estate firms, which are driving this business. We smallholders are only spectators in this development.*"

An elderly smallholder from Huancayo

Residential development at the peri-urban interface of Huancayo leads to the loss of irrigated farmland (A. Haller)

the valley cannot be grown in these belts owing to lower temperatures. Moreover, while they practise year-round potato production in the valley, it is not possible in the slopes and high plains because they are not irrigated.

Farmers have thus turned to other solutions, planting the steep and non-irrigated lower slopes with eucalyptus trees, as timber is in high demand by the urban construction sector. Higher upslope, terraces make their appearance for seasonal production of potatoes and other tubers such as oca (*Oxalis tuberosa*), olluco (*Ullucus tuberosus*) and mashua (*Tropaeolum tuberosum*). Moreover, the nearby puna grasslands are undergoing a major land cover change due to burning and pastoral expansion (4). Since urban policy-makers consider range burning a driver of global warming, regional deglaciation and soil erosion, and thus a reason for local water scarcity, they have established a regional conservation area (6) to regulate land use in the grasslands. This includes a plan to substitute sheep and cattle with alpacas (7).

In sum, the peri-urban smallholders of Huancayo understand the challenges and opportunities of urban growth and hope to profit from a growing urban market. They have developed new ways to generate income at different altitudes as an alternative to the ground lost in the valley. However, in order to create a flourishing rural–urban interface, they need the support of planners and policy-makers, especially relating to strengthening smallholder-market linkages.

Lessons learned

- In developing mountain regions, the rapid growth of cities is no longer exclusively an urban concern. Studies, assessments and policy measures made for urban development should therefore take into account the agrarian hinterlands and the adaptation strategies developed by rural communities.

- Due to socio-economic disparities in the peri-urban interface, it is crucial to consider the different stakeholders' perceptions of the impact of urban growth. Peri-urban decision-making should hence be based on multiple criteria that also include the local agriculturalists' assessments, in order to prevent land use conflicts and negative effects on smallholder food and income security.

Urban sprawling increases land prices in the fertile valley (A. Haller)

Zone	Altitude	Land tenure	Land use before urbanization	Land use adapted to urbanization
puna	4,000–4,800 m	State-owned; agrarian communities hold land use rights shared between members	Extensive production of sheep and cattle all year round	Intensive production of sheep and cattle all year round
				Rainfed cultivation of potatoes, mashua, oca and olluco
suni	3,500–4,000 m		Rainfed cultivation of potatoes, mashua, oca and olluco	Production of wood crops (*Eucalyptus* spp.) all year round
Quechua	3,300–3,500 m	Private-owned; few non-agrarian big landowners and some agrarian minifundistas	Irrigated cultivation of potatoes, maize and artichokes	Residential urban; intensive breeding of small animals (guinea pigs) all year round

Table 1: Strategies of adaption to urbanization taking the example of the Shullcas Valley, Huancayo

3 Learning and cooperation

Women cooking together for the community's patron saint festivities, Peru (S.-L. Mathez-Stiefel)

Over one hundred people from the private and public sectors discuss strategies to involve private landowners, ranchers and farmers in cooperative conservation in Seeley Lake, Montana, USA (A. Duvall)

Jill M. Belsky

Learning and cooperation

Family farmers in mountain areas historically cooperated with each other to earn their living and steward natural resources. Today the forms of cooperative action involve new and innovative partnerships and collaborations across increasingly diverse types of people, sectors and enterprises.

Working cooperatively has long helped mountain communities deal with harsh climates, remote locations and labour-intensive livelihoods. Cooperation underscores centuries of common property rights and governance in mountain forests, farms and grazing lands (1). With the current focus on sustainable mountain development, these communities are finding that the rapid transformations in economy, ecology and demography in mountain areas create both new opportunities and challenges for learning and cooperation.

Greater access through roads and markets as well as Internet and other technologies (including radio and television) has greatly expanded communication potential in mountain areas. This, in turn, has provided a critical means for mountain people and groups to access information for economic development as well as to bolster their local identities.

Partnerships have become a key mechanism for pursuing new livelihoods in mountain areas or adding value to existing ones. This happens, for example, when local agro-food producers form a cooperative that teams up with a large supermarket chain to brand and market its goods. Bundling resources and multi-sector approaches are particularly adaptive in mountain contexts. Partnerships between private and public entities are particularly vital in scaling up local efforts to make meaningful impacts, or when public funding is unavailable.

However, a focus on partnerships and cooperation also creates challenges. Collaboration brings different groups with different interests together to work for a mutually agreed-upon outcome, but there are often unequal conditions upon which individuals and groups collaborate. Hence there is a likelihood that some interests will dominate over others, which makes it particularly relevant for learning or applying democratic principles and practices.

Groups with similar interests and histories of cooperation are particularly well positioned to cooperate and succeed. Women's groups suggest the value of gender-specific arrangements and building on their traditional cooperative practices.

Mountain areas offer many examples of collaboration and cooperation being used deliberately to reconcile diverse interests. For example, in the USA, collaborative efforts are increasingly being used to determine how forests, ranchlands and aquatic resources in mountain ecosystems are owned, restored, stewarded and managed for diverse ecological, economic and cultural values (Box). Despite real and enduring challenges to cooperative action, examples from around the world suggest that it remains a key process in the sustainable development of mountain regions and resources, and that with proper recognition and support, it can become an even more critical resource in the future.

A key recommendation is to support cooperative efforts for sustainable mountain development. This involves recognizing the diverse ways in which mountain people and groups are pursuing new enterprises with new partners within mountain areas and beyond. Public and private partnerships need to be encouraged. Valuing and building on historical knowledge and practice will also ensure that new ventures have local meaning and connection. Recognizing the strengths of community-based initiatives and enterprises where they are working successfully can also strengthen sustainable mountain development. Lastly, avenues for information-sharing and communication that specifically include mountain people and mountain places should be promoted.

Cooperation conserves forest and ranchland across large landscapes in Montana, USA

When a global timber company, Plum Creek, announced it was planning to sell thousands of hectares of former timberlands in Montana, a diverse group of people became concerned. In the absence of government regulations, these lands would likely be subdivided and converted into vacation homes, reducing resources for local livelihoods in forestry and ranching, public recreation and for protecting wildlife habitat and other ecological services. Under the Blackfoot Community Project and Montana Legacy Project, private citizens, community-based organizations, business leaders, sportspeople, elected officials, federal and state agencies, conservation groups, and university personnel joined forces to purchase approximately 160,000 hectares of former Plum Creek timberlands throughout the Northern Rocky Mountains/Crown of the Continent ecosystem. Because of their cooperation these lands now have a mixture of public and private ownership with mandates to restore and sustainably manage them to enhance their ecological, cultural and economic values [2].

Faena (collective work) to repair and clean the community's irrigation channel in Pitumarca, Peru
(S.-L. Mathez-Stiefel)

Building on traditional cooperation among women

Paolo Ceci, Fatoumata Binta Sombily Diallo, Petra Wolter, Lavinia Monforte

In the Fouta Djallon Highlands in West Africa, solidarity and collaboration among women have traditionally ensured they can rely on mutual assistance in case of need. Building on these practices, development projects have established women's interest groups in the area, with the aim of increasing and diversifying incomes of small family farms.

Women tilling a kitchen garden for sowing (P. Ceci)

The Fouta Djallon Highlands are a series of plateaus ranging from 900 to 1,500 m altitude in the central part of Guinea, extending into Guinea-Bissau, Mali, Senegal and Sierra Leone. They are the area of origin of important rivers including the Gambia, Niger, Senegal and Konkouré Rivers. Subsistence agriculture, based on small-scale family farming, remains the principal source of livelihoods. Due to the mountainous topography of the area, crops are produced in the valley bottoms, the plains and on steep slopes, but also in kitchen gardens, which are cultivated exclusively by women.

Women play a crucial role in agriculture, livestock breeding, family nutrition and health care but also in domestic tasks such as meal preparation, water and fuelwood collection, house cleaning and laundry, usually assisted by daughters or granddaughters (Table 1). However, they are also increasingly taking on traditionally male duties, as men leave the rural areas in search of work elsewhere (1,2). In addition to these tremendous workloads, rural women are also disadvantaged in other ways. They have lower education levels than men, which affects their ability to access information, agricultural extension services and farm inputs, including improved technologies. Where ownership or usage of land is concerned, men habitually claim priority and hereditary rights.

However, due to the fact that women readily accept working collaboratively (Box), development projects have supported the establishment and legal recognition of women's groups since the 1980s. These projects serve to strengthen women's positions, improve agricultural production, and increase and diversify household incomes. The focus on vegetable production in the fertile valley bottoms has boosted production of cabbages, tomatoes, aubergines, chilli peppers, lettuce and spinach, among other products. Thanks to these efforts, women have both

Women of a market garden group and villagers in Fello Ferobhé, Bantignel (P. Ceci)

Lessons learned

- The strength of the interest group approach lies in its potential to reach a larger number of people for training and exchange. It builds on motivation, interest and commitment expressed by local actors such as women who join forces to pursue common objectives.

- Experience shows that entrepreneurial skills are equally important to diversify group activities and develop innovative and profitable ventures. There is need for leadership and empowering women to take on decision-making and increase their negotiation capacities, especially when it comes to marketing.

enriched their family diets and become more financially independent and, in turn, better able to contribute to school fees of their children, which has resulted in increased enrolment rates. However, the increased production supply of vegetables has also meant substantial seasonal price drops, forcing women to sell their hard-earned produce below cost.

In order to remedy this situation, FAO has engaged in a project to improve vegetable production, post-harvest handling, storage, processing and marketing. It targets individual farmers with entrepreneurial aspirations as well as women's groups (3) – testing sustainable conservation and transformation techniques for surplus agricultural production, and enhancing the capacities of women in business development. The project includes labelling high-value processed products from organic agriculture and native trees, such as shea butter, and developing market networks and value chains that also include poorer households. It also identifies links between nutrition and health by, for example, testing the result of using locally produced groundnut oil to replace imported and less healthy palm oil. Governance issues are also on the project agenda as men tend to occupy managerial positions even in women's groups – including the presidency, secretariat and bookkeeping – resulting in an imbalance in decision-making.

Woman watering tomatoes in the market garden (P. Ceci)

Traditional forms of women's mutual assistance, Fouta Djallon Highlands, Guinea

Several forms of mutual assistance exist among Fula women in the Fouta Djallon Highlands, Guinea: The most widespread is *Kilé*, which confers on women the right and privilege to invite their entire village to work on steep slopes. The villagers go with their own tools to the field of the convener, who provides a rice and meat lunch (4). However, Kilé arrangements are costly and, therefore, restricted to the richer social strata. A smaller-scale version of *Kilé*, called *Kilé Futu*, is convened by older women, who periodically gather their young nephews and nieces to help out.

Other forms of mutual assistance include *Ballal*, which appeals to family solidarity, for instance to all women living in the same family compound. There is also *Yirdè*, associations of youth of the same age that provide assistance in exchange for meals or small compensations. Under the arrangement of *Tontine*, women periodically collect money and each takes a turn as recipient. Remunerated work can also be a solution for those who have the means to hire daily labourers. More recently, women who head a household but lack sufficient economic resources to employ labour, started joining forces and reciprocally assisting each other in heavy agricultural tasks formerly carried out by men, such as ploughing, sowing, harvesting and threshing.

Key feature of family farms (N=95)	Percentage of farms
Agriculture as main source of livelihood	75%
Affected by male outmigration	71%
Female-headed	24%
Women contribute to children's educational cost	38%
Women contribute to family health care expenses	46%
Women are members of interest groups	37%

Table 1: Key features of family farms, Guetoya, Prefecture of Pita, Fouta Djallon Highlands, Guinea (5)

A farmers' cooperative and a supermarket team up

Markus Schermer and Christoph Furtschegger

For the past ten years the Bio vom Berg (organics from the mountains) brand has been an inspiring example of how organic products from a mountain region can be successfully marketed. The cooperative Bioalpin, which unites organic farmers, small-scale processors and permanent members, owns the Bio vom Berg brand. In 2002 Bioalpin joined forces with a supermarket chain, and today, some 600 farmers benefit from the initiative as suppliers.

Promotion campaign of an organic product, Austria (©MPreis)

It all started as a lucky coincidence. When the organic farmers in the Austrian state of Tyrol wanted to establish a trading platform to collect their produce and sell it collectively to retailers, they received assistance from the regional agricultural marketing board, Agrarmarketing Tyrol, which subsidized the initial personnel costs. At the same time, a regional, family-owned supermarket chain, MPreis, wanted to improve its profile by building a brand for organic products, believing that a producer-owned brand, rather than a retail brand, would increase consumer trust. Thus, it teamed up with the Bioalpin cooperative, which created an identity for its Bio vom Berg brand with the slogan *"delivering the best products from local organic farmers to the food store around the corner"*.

MPreis operates more than 200 stores within Tyrol and adjacent areas and is thus a strong partner for the cooperative. It works closely with the cooperative on product development and marketing, and has increased its initial 8 products to over 80. The cooperative includes 30 processors, among them 10 local organic dairies and one organic butcher; producer associations for eggs, fruits, grains and potatoes; and individual farmers who specialize in vegetables or berries. Altogether, about 600 out of a total of 3,000 organic farms benefit from the initiative as members or as suppliers. The turnover increased from EUR 672,000 in 2003 to around EUR 5 million in 2011, with the cooperative paying fair prices, in order to help preserve small-scale structures in producing and processing.

"*In a controlled organic and local production, I think that the most sensible way to produce is in accordance with traditional values of food of invaluable quality. With our work we sustain small-scale Tyrolean mountain farms for future generations and provide valuable, natural products from the region.*"

Heinz Gstir, the chairman of Bioalpin, defining his vision

Different partners joined together for the success of Bioalpin (©MPreis)

Lessons learned

- A collective marketing initiative for small-scale farmers and processors can be successful if they retain the power over their resources and team up with committed large-scale commercial partners. Being regional and organic provides a collectively shared notion of quality, essential for establishing long-term relations of trust with clients and consumers.
- The collective approach allows improved coordination and the necessary specialization within small-scale structures, while also retaining a broad product range.
- While strong partners are critical for success, ownership of the brand as well as diversification of marketing channels is essential for maintaining independence on the producers' side.

The main asset of the cooperative is its ownership of the Bio vom Berg brand. Right from the start, the founders wanted to establish this brand in contrast to other existing organic retail brands. As the cooperative owns the brand, it has a stronger position in marketing and price negotiations: retailers cannot easily switch suppliers in order to undercut prices as the farmers – who are all members of the cooperative – are the owners of the brand.

While the close relationship with a strong retail partner has a number of advantages, it still presents potential pitfalls, such as difficulty in maintaining independence. Therefore the cooperative has started a number of initiatives and projects with other partners. For example, a special grain project, based on traditional local varieties, supplies the biggest Tyrolean bakery, which sells organic bread under the Bio vom Berg brand in 70 outlets throughout the state. More recently, it has begun to supply products to regional hotels and restaurants, an ambitious goal that has a great potential in a key tourist region like Tyrol.

Bioalpin's approach and philosophy

Bioalpin acts as a trading platform among farmers, processing enterprises and retailers. It coordinates production, negotiates price and quantity with its purchasing partners and organizes logistics. The organization is kept rather small. The main goal is to organize, coordinate and synchronize individual farmers within producer groups. This helps reduce the number of contact people and improves the personal relationship among partners.

The philosophy is based on an alternative concept of growth. Instead of the usual growth per farm unit, network growth is propagated: While the number of farms involved is constantly growing, each farm can still maintain the positive features of a small structure and specialize in part of its production. The establishment of producer groups allows internal coordination and exchange, and helps keep the costs for the cooperative minimal. The bundling of products in terms of variety and quantity increases the cooperative's bargaining power and thus helps secure reasonable prices for primary producers.

Radio Mampita – the powerful voice of rural people

Felicitas Bachmann

In Madagascar, where about 80% of the population is rural, access to timely and reliable information and to the services of governmental and non-governmental organizations has been a major challenge. Now, rural communities have been connected by an independent farmer-owned radio station, Radio Mampita, which also supports knowledge and information exchange, and renders rural development more demand-driven.

A local correspondent interviews members of a women's group, Madagascar (E. Gabathuler)

Radio Mampita in the Haute Matsiatra region, one of the few rural-based Malagasy broadcasting stations, aims to empower rural communities by enhancing rural communication and giving rural people a voice. These people have been isolated owing to difficult topography and limited road access. They have lacked access to information and means of communication.

Before launching Radio Mampita in 1997, the Swiss Agency for Development and Cooperation (SDC) invested several years building the necessary skills and a functioning organizational structure to ensure continuity. This began by providing five journalists with multi-media training to develop skills for communicating with, and informing about, the issues of rural people. Next, an awareness-raising campaign was initiated about the possibility of having a farmer-owned radio station and the need to establish a farmer association that could take over the ownership. Finally, villagers who volunteered to act as local correspondents in their communities were trained in collecting information, conducting interviews and facilitating public debates, and in sending the registered material to the radio station, where the broadcasts are produced and broadcast. In 1997, the Association Mampita, consisting of farmer organizations, was established as a non-commercial, politically independent and religiously neutral institution (Figure 1).

Broadcasting from the city of Fianarantsoa and covering a perimeter of 70 km in the Haute Matsiatra area, Radio Mampita reaches approximately 1 million people. Initially, fully dependent on donor funding, it steadily increased its revenues until it became financially independent in 2007. The sale of broadcasting time to institutional partners generated 30% of the revenues in 2010, while airing of personal messages and announcements amounted to 70%. Today, Radio Mampita

"*We*, the female artisans, regularly contribute to Radio Mampita's programme called 'What about us, the women?'. That's why we and our products are widely known and we recently received a lot of orders. In addition, we are often invited to present and exhibit our handicrafts at national fairs."

Joséphine, a local artisan

![Women listening to the radio programme 'What about us, the women?' (E. Gabathuler)]

addresses the entire rural population, i.e. men and women, adults and children, and covers issues and debates with an educational or information focus (13% of broadcasting time) including health, agriculture, market information and civil rights; news (36%) including news from the villages and announcements from service providers or of family events; and entertainment (51%) such as music, radio plays and greetings.

A study conducted in 2010/2011 credited Radio Mampita with a number of changes in the area (1):
- rural people's access to relevant information, e.g. on agricultural techniques, laws and civil rights, etc., had strongly improved
- Radio Mampita had become a widely recognized mouthpiece of the rural population
- the negotiation power of producer organizations had improved and rural economies were supported through better access to timely market information and by linking producers and buyers, thus eliminating intermediaries
- rural people's interaction with service providers had become more self-confident and proactive, resulting in a positive competition among development organizations and more demand-driven support activities
- communication among communities and family members had become much easier and cheaper
- security had improved as, in the case of a criminal act, a message over radio enabled fast reactions to unusual incidents.

Lessons learned

Radio Mampita plays a dynamic role in Haute Matsiatra and is highly appreciated by both its target audience and actors in rural development. Key factors of success are the following.

- Its identity as the farmers' radio owned by farmer organizations and strongly anchored in the rural world. It strictly broadcasts in the local dialect (Betsileo), and 90% of all programmes are directly related to the rural world.
- Its political and religious independence is very much appreciated by development actors. Due to its strict neutrality, Radio Mampita has survived several political crises.
- Its local correspondents, being villagers themselves, know exactly the needs and concerns of the rural people in the highland.
- Its manager and staff, being themselves of rural origin, are highly motivated and committed to the mission of Radio Mampita.

Farmer in his rice field (E. Gabathuler)

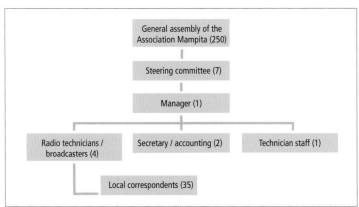

Figure 1: The organizational set-up of Radio Mampita (brackets: number of members) (1)

A school for promoters of agro-ecology

Francisco Medina and Jenny Chimacyo

In the Apurimac region of the Peruvian Andes, food security depends on the availability of fertile soils. Yet, 92% of the land is extremely vulnerable to erosion and desertification and only 10% is suitable for farming. As increasing drought conditions have provoked high losses to the 2,500 smallholder family farmers living in 23 peasant communities, a training programme has turned their farms into classrooms to improve their livelihoods.

Capacity building of local actors in the Apurimac region, Peru (© A. Escalante)

Apurimac is dominated by peasant communities that practise subsistence farming, including some animal production based on alpacas, llamas and vicuñas. Average monthly income ranges from USD 70 to 215 and its Human Development Index (HDI) is quite low, ranging between 0.49 and 0.56. Its farm economy is characterized by the predominance of family labour with production that is mainly for household consumption although occasionally for sale at the local market. There is a lack of agricultural support services, and scarce or no processing of agricultural raw materials.

It was against this background that a training programme – the School for Agro-ecology Promoters – was created in 2011. Its aim was to improve the water, soil and agro-biodiversity management, in a bid to support farmers in managing their land sustainably at the household level as well as the level of the watersheds. The classrooms are the farms themselves. Trainers visit twice a month to advise farmers in resource management based on a set of practices that include traditional as well as new knowledge. This promotes recovery and improvement of traditional systems such as slow-forming terraces bordered by indigenous vegetation, infiltration trenches and systematic crop rotation. Farmers learn to improve seed selection and irrigation and to upgrade soil quality by using organic fertilizers such as compost. They also use foliar fertilizer produced from alfalfa and other plants for pest control and mineral mixtures such as copper sulfate or quicklime for fungal control. Courses are followed up by project technicians who visit the farmers periodically, and the project sets up opportunities for farmers to intern in other communities known for good agro-ecological practices.

The three-year study cycle covers soil management and conservation techniques during the first year, and moves on to negotiation processes, marketing and organization

"*The key is rescuing ancestral knowledge, all that knowledge that existed and that was being lost, and combining it with scientific techniques.*"

Eugenio Paúcar, farmer and promoter

"*I have everything here. It comes directly from the farm to the table. There is nothing better than eating what we produce ourselves. It tastes much better.*"

Griselda Letona, farmer and promoter

Farmers at harvest of wheat (© MST Apurimac/MINAM/PNUD/GEF)

Lessons learned

- Following a community approach is important, as only a community with a strong social cohesion will be able to reach the agreements and regulations required for sustainable management of its natural resource base.
- The concrete experiences and lessons learned by the family farms engaged in the project will inform policy-making – and contribute to the formulation of the national policy for sustainable land management, which the Peruvian Ministry of Environment is in the process of formulating. The institutional link is important: The ministry is the implementing agency of the project.

The planting stick helps farmers master the land in the Andes (© A. Escalante)

and management of food security in the second and third years. Courses follow the agricultural cycle, starting when the harvest ends in May and ending in April of the following year, so that farmers can apply what they learn in real time. Each month, trained farmers share the knowledge they have acquired in communal meetings, following the "farmer-to-farmer" approach. Thus, "students" from the first cycle become trainers in the subsequent cycles.

The project life is five years (2010–2015) and covers 2,500 households. During the first year, the project trained 95 promoters, 27 of whom were women, and reached 405 families. In the second year, another 90 promoters were trained, increasing outreach to 675 families. During the 2012–2013 training campaign, participants planted 651 ha of native crops, and yields increased between 150 and 250%. In 2012, farmers established a producers' association to facilitate fair trade of products, which now includes more than 800 families. The association is the last step in a process of strengthening community organizations so that they are able to plan, propose and evaluate initiatives that promote the use of natural resources without compromising their regenerative capacity.

Practices encouraged by the Promoters' School:
- undertake germplasm management and in-situ conservation of a diversity of plant genetic resources, including potatoes, maize, quinoa and diverse fruits
- use crop diversification and mixed cropping with leguminous plants that fix nitrogen and improve soil quality (use of organic fertilizer from compost)
- increase fodder crop cover, including alfalfa, ryegrass and oats, to reduce soil erosion and increase fodder availability, coupled with hay production and silage
- use pressurized irrigation systems for efficient use of water in water-scarce periods
- measure the yield of springs for assessing water availability for the expansion of irrigation
- improve shearing alpaca and the selection and packaging of yarn.

Field schools for agro-pastoralists

Caterina Batello, James Okoth, Monica Petri, Manuela Allara

Pastoralists and their families fulfil an important, but largely unrecognized role in agriculture production and natural resource management in many mountain and upland regions of the world. Pastoralists are among the most underprivileged and marginalized groups in today's sedentary world. Programmes such as Farmers Field Schools show the value and development potential of the pastoral way of life.

Cattle in a protected *kraal* (enclosure)

At a global scale, approximately 1 billion people depend on pastoral production of livestock, which serves as a source of income and food security for 70% of the world's 880 million rural poor who live on less than USD 1 per day (1). In the Horn of Africa, for example, the volume of informal livestock trade is estimated at more than USD 1 billion per year, and in East Africa, 56% of the Nile basin is used by pastoralists (2).

Lamentably, mainstream development narratives perceive pastoralism as a backward and wasteful lifestyle that causes degradation of rangeland and grasslands and creates conflicts with non-pastoral people. In short, pastoralists are viewed as generating low profits, uneducated, archaic, poor and destined to disappear (Box).

To counter this, many programmes that recognize the important role of pastoral communities are promoting activities in support of these communities as central actors of food security and an integral part of healthy social-ecological systems, especially in drier mountains and upland regions. For example, the Karamoja Region of Uganda, which has 20% of the country's cattle and almost 50% of its sheep population, is one of the most vulnerable regions of the country owing to climate variability, drought and transboundary livestock diseases. Thus, Farmers Field School programmes developed with the support of FAO for the area focus on the capacity of pastoralists to manage, restore and protect natural resources while producing meat, milk and other food, and enhancing the capacity to diversify revenue.

A Farmers Field School programme starts with joint exploration of the main issues affecting pastoral households and develops a curriculum to address these issues. Partners are brought in to identify specific technologies and practices for testing. Promising measures are then implemented in a pilot programme, and a review involving pastoralists, facilitators and project partners evaluates which of these activities could be upscaled to the larger community.

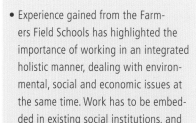
Agro-pastoralists map the local resources

Lessons learned

- Experience gained from the Farmers Field Schools has highlighted the importance of working in an integrated holistic manner, dealing with environmental, social and economic issues at the same time. Work has to be embedded in existing social institutions, and local communities are at the centre of any action that intends to generate lasting solutions.

- Globally, approximately one billion people depend on pastoralism and agro-pastoralism, and many of them are in extreme poverty. Policies and programmes targeted at food security and poverty reduction should therefore have pastoralists at centre stage.

- Investments in a science, infrastructure, and education relevant to pastoralists, and in the development of technologies adapted to their needs, should be supported much stronger than it is the case at present.

Milk helps improve children's diet

The Farmers Field School programmes are based on crop–livestock production and land and water management, including disaster and risk management, and a holistic catchment-based approach. The programme has developed community action plans jointly with the pastoralists and agro-pastoralists in order to develop measures that minimize the effects of climate variability on livelihoods. The programme also introduces sustainable crop production intensification, community animal health, natural resource management and alternative revenue generation. It promotes revitalization of local landraces, in recognition of their potential for increasing resilience against the vagaries of climate. The schools provide animal disease surveillance and diagnostic services complementing the dramatic shortage of veterinary services in the region. In a bid to improve animal nutrition and to increase health and resistance of livestock, forage legume trees have been planted, and grasslands oversown with legumes to improve their nutritional value for livestock. Vegetable production and beekeeping have been introduced as alternative livelihood sources.

Pastoralists as portrayed in the media

A media review on pastoralism in Kenya, India and China revealed that pastoralists' voices and opinions were included in less than a half of the 170 articles analysed, and the voice of pastoral women in only 21 of them. Only 6 articles out of the 170 praise mobility as a sustainable management practice in drylands. Only few refer to ways in which pastoralism can contribute to food security [3].

Lobbying for mountain regions and farming

Jörg Beck

The importance of mountain regions in Switzerland has decreased as urbanization has increased. The urban population often only views mountain areas as landscapes and wilderness, and fails to understand the needs of mountain family farmers. The Swiss Centre for Mountain Regions (SCMR) has shown how constructive influence can help align the interests of mountain areas with federal policies and projects.

SCMR brings the concerns of mountain people to the authorities in the capital (SCMR)

Tourism as well as society in general benefit from well-maintained landscapes, yet the traditional landscape of mountain regions cannot be maintained without agriculture. Recognizing this, the 2014–17 Swiss agricultural policy aims in the right direction by offering better compensation for the public services provided by agriculture (1). However, the focus of a strategy for mountain regions must be multi-sectoral, and the framework must allow the self-determined development of these regions.

Since its founding in 1943, SCMR has lobbied for family farms, recognizing them as key pillars of rural development in mountains (Box). SCMR has worked to improve living conditions and development opportunities by defending the economic, political and cultural interests of mountain people, but also by coordinating various local, regional, cantonal, national and sectoral efforts to promote mountain areas in Switzerland and abroad.

Organized as an association and governed by a General Assembly (2), SCMR has both public-law and private-law collective and individual members. It represents 23 cantons, 700 municipalities, 30 tourism and some 100 agricultural and commercial organizations. The "Rat der Berggebiete" (Council of Mountain Areas) is elected by the General Assembly and has a well-balanced representation of national regions and of the different sectors. It meets once a year and advises the managing board regarding issues of strategic importance. The managing board decides on daily business and advises various actors on political initiatives and statements.

In order to achieve its overall goal, SCMR develops activities meant to:
- influence politics relevant for mountain areas
- inform the public and political decision-makers on mountain-related issues
- promote education and research in and about the mountain area
- take practical measures and support projects in the mountain area for the benefits of all stakeholders.

Mechanized farming in mountain areas is limited, and the benefits are low (V. Gilloz)

Lessons learned

- SCMR's support is key for the development of effective local and regional networks and organizations. Thanks to its wide and diversified network, SCMR has become one of the crucial players for the regional development in Switzerland and abroad.
- Using a holistic, multi-sectoral approach has been an effective strategy in striving for democratically proven solutions that favour mountain regions.
- Close contact to the local level ensures two-way communication. This allows introducing relevant issues into the political debate.

These practical actions are manifold and include providing technical advice on issues of regional development and mountain agriculture, facilitating processes and development of cooperation models, supporting the construction of rural buildings for cooperatives, and brokering of voluntary services and work in mountain areas.

In order to advance the multi-sectoral development in mountains and thus maintain mountain regions as a place for living, working and leisure, SCMR has undertaken several successful initiatives, including:
- conducting a study in 2008 (3), which found that bundling of offers and services for tourists could boost agrotourism in mountains (which is now supported by the federal government)
- calling for better recognition of part-time farming in the context of the new Swiss agricultural policy
- launching the brand "Swiss mountains" in 1995 for the protection of the term "mountain", with the federal government going even further by establishing a mountain and alp ordinance (BAlV), which contributes to protecting the terms "mountain" and "alp" to compensate the locational disadvantage (4) (this regulation will be adopted by the European Union)
- campaigning with success to establish broadband access as a service of general interest in the federal public service mission for the telecommunication sector – under the guideline "Towards the Information Superhighway" SCMR supported the expansion of broadband technology (5) to mountain areas, allowing mountain people and family farms to keep pace with technology development that offers new opportunities.

Family farming: the fundament of decentralized settlement

Agriculture makes an important contribution to decentralized, countrywide land use. Family farms have a key function in this. Out of approximately 57,000 farms in Switzerland, about one-third is located in the mountain area (6). Due to the accelerated structural change in recent years, this number is decreasing steadily. Agricultural income in mountain areas is highly dependent on direct federal payments. Markets are distant and product prices are under strong pressure. Due to the topographic and climatic conditions, mountain agriculture is only capable of reacting to market changes to a limited extent. Agricultural income in the mountain areas is traditionally low and amounts to only approximately 60% of the income of a farmer in the plains (6), making it important for family farms in mountain areas to have additional income from outside the agricultural sector.

4

Sustainable intensification and organic farming

Kitchen gardens in Kara-Teit village, Kyrgyzstan, help improve children's health (A. Abazbekov kyzy)

Lao farmers improve the cultivation of *Khao Khai Noi*, a local rice variety (C. Flint)

Maria Wurzinger and Urs Niggli

Sustainable intensification and organic farming

Sustainable intensification has been defined as a form of production wherein yields are increased without adverse environmental impact and without the cultivation of more land (1). However, being able to achieve sustainable intensification depends on the context but also on the views of the different stakeholders (2).

While "sustainable intensification" is a term widely used today, other more descriptive terms have also been added to the lexicon, such as "ecological intensification" and especially "eco-functional intensification" as used in the Research Vision for 2020 of the International Federation of Organic Agriculture Movements (IFOAM) (3). Eco-functional intensification refers to productivity increases that may result from ecosystem functions and services such as fertile soils, diversified landscapes and fields as well as from diversified production activities that make farms more resilient, both ecologically and economically.

In contrast to the intensive farming landscapes of the plains, mountain regions are hotspots of biological diversity. Farmers and pastoralists in these regions often live and work under extreme and harsh conditions, facing climatic stress and poor soil fertility, but also poor infrastructure and lack of market access – all of which limit their options to intensify their farming practices. The mixed crop–livestock systems or solely livestock systems found in mountain regions present different leverage points for interventions.

In mountain regions of Europe, organic family farms are often the most viable form of agriculture. Where the access to markets is well organized, they are even the preponderant form such as in some regions of Austria and Switzerland (4). Many families practise livestock-rearing, dairy production or mixed farming. An improvement or intensification in livestock production can only be achieved by taking a production system perspective and addressing different problematic areas at the same time. Improved, but well-adapted breeds, better pasture management and improved animal health are key factors in this intensification process. Species-rich grasslands with positive effects on animal fertility, longevity and milk quality offer a good example of how botanical diversity can be used for productivity gains.

Organic family farms – both certified for remote markets and uncertified for local ones – also have grown in number in Latin America and Africa, especially in mountain regions. An analysis of thousands of farms (5) concluded that the better use of nature, human and social capital of organic farmers led to productivity increases of more than 100% and improved livelihoods, which, in turn, provided for higher education for farm children. The study highlighted the positive effects on sustainable ecosystem management due to increased training of, and improved cooperation among, farmers.

Sheep farmers in the Ethiopian highlands (M. Wurzinger)

Intensification is more likely to happen in an ecological way when farmers have access to markets where there is a demand for their products. In many developing countries, there is an ongoing urbanization process, which includes mountain regions and offers new opportunities for farmers. Niche markets will be becoming more important in Europe and might offer new opportunities for products from mountain regions.

Sustainable intensification at farm level can best be achieved if there are a support system and policies in place that compensate farmers for the provision of public goods. This requires recognizing the important role of rural families in remote mountain regions for society. Their farming practices are relevant for the conservation and protection of common goods such as freshwater, and they are important custodians of biodiversity.

Innovations, not only technical but also organizational and institutional ones, support farmers in their daily endeavours to make a living under extreme conditions. In addition, various types of ecosystem service markets can contribute to integrating both livelihoods and ecological objectives, such as carbon markets, in the near future.

Organic certification is another form of ecosystem service payments – as it allows farmers to benefit from price premiums for their sustainable and healthy farming practices. Achieving certification includes more than farmers having training in best farm practice and farming with little or no use of agrochemicals – they also have knowledge of quality control and market access.

The potential to increase productivity by using agro-ecological methods is relevant and should be promoted by support schemes and policies, especially as farmers in mountain regions play an important role in a multifunctional agriculture. Certification is often an important tool for sustainable intensification and an additional way to compensate for the provision of public goods.

Towards a fully organic state

Niraj Nirola and Trilochan Pandey

The Indian state of Sikkim is a biodiversity hotspot located in the Himalayas, with 47% forest land and only 12.3% farmed land. Dynamic, adaptive and diverse, Sikkim has adopted a statewide policy to convert its low-input-based family farming system to organic farming. However, although the state's organic effort is unique, experience suggests the need for a more holistic and sensitive policy approach.

Women tending to the potato crop in Ribdi (N. Nirola)

In Sikkim, where 64% of the population is directly dependent on family farming, the agriculture sector faced sluggish or even negative growth for several years. This prompted the government to adopt a statewide organic farming policy – a policy meant to make farming more profitable by bringing premium prices for organic products, creating jobs and increasing self-reliance, but also to help preserve the fragile mountain environment.

Once the Sikkim State Organic Board was established in 2003, the state initiated a ban on synthetic inputs and by 2006 had completely revoked central fertilizer subsidies. The Sikkim Organic Mission was launched in 2010 to implement and monitor the Organic Board's programme to convert 50,000 ha into certified organic farming land by 2015. The government bears the costs of certification and provides free inputs, equipment, training and extension services. It also promotes the use of bio-fertilizers and organic manure. Certification mechanisms are instituted through accredited certification agencies. An Internal Control System (ICS) – which acts as the intermediary between farmers, government institutions and certification agencies – has outsourced its work to 13 private companies, which are paid for successful conversion on a per hectare basis (Figure 1).

Organic farming requires collaboration with private sector and across sectors

- 3 accredited global and national certification agencies are responsible for the certification of farms and products
- 13 private companies have the task to control the implementation of organic farming
- 5 governmental organizations support the transformation to organic farming.

The traditional and low-input subsistence character of family farming has helped smooth implementation of organic farming, although with some resistance due to farmers' attitudes about changing practices (Table 1). For example, Budang village has been fully certified for three years, but farmers still face difficulties in producing sufficient manure, due to decreasing availability of fodder and livestock. Sri Badam has had an easier time adopting organic farming owing to the village's dependence on a large, agroforestry-based cardamom cultivation and its relatively larger livestock population. But in Ribdi village, which relies on high-input potato farming and has a low livestock population and low manure production, the farmers openly use synthetic inputs and resist the organic initiative. To compensate for expected yield loss, Ribdi farmers are asking for money or government jobs, but the government has no compensation plan nor does it have the ability to offer farmers government jobs.

Villages surveyed	Budang (N=15)	Sri Badam (N=16)	Ribdi (N=12)
Altitude	500 m	2,100 m	3,000 m
Total population*	2,488	984	915
Percentage of female population	51%	53%	49%
Percentage of households below poverty line	40%	93.75%	75%
Farming technology	Organic (certified)	Organic (in conversion)	Chemical-based farming
Major crops cultivated	Rice, maize, ginger, mustard, buckwheat, tapioca	Maize, cardamom, pea, potato	Potato, pea, cabbage
Farms self-sufficient in farm manure	46%	75%	no answer

Table 1: Villages adopt organic farming to different degrees based on their specific socio-economic situation. *Source: Census of India 2011

The organic farming initiative is facing various challenges. The market for effective commercial bio-pesticides and bio-fertilizers is undeveloped. Farmers resort to short-term measures such as continued use of synthetic inputs from outside the state. Because the revenue of ICS providers is linked only to how many hectares are converted, they have an incentive to overlook such malpractices. Moreover, some traders sell their non-organic products but use the organic tag. The farmers are dissatisfied with previous marketing efforts for cash crops and with the fact that, thus far, organic products do not bring higher prices than conventional products. Women farmers have difficulty finding sites to sell their products in the weekly local markets and compete with vendors who sell at lower prices. In the future, Sikkim's farmers will need improved marketing and distribution schemes for their organic products in both the domestic and export markets.

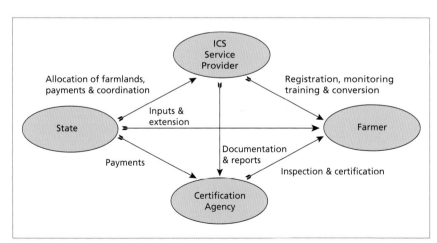

Figure 1: Model of organic farming in Sikkim

Lessons learned

- The state should work towards promoting and conserving traditional farming systems, which can provide a niche for organic farming.
- The "one size fits all" approach overlooks the specific needs, and the diversity of agronomic, ecological and socio-economic conditions across the state. A sense of ownership, involvement and stakes among the farmers is yet to be created.
- Emphasis on large-scale and total certification neglects the need for developing market linkages and building farmers' capacity. It is uncertain whether farmers will be ready to bear the cost of the annual renewal of certification in future.
- Introduction of organic farming in stages and on a smaller scale that is supported by infrastructure development and market linkages with farmers' groups and cooperatives will increase the farmers' stakes and sense of ownership.

Women selling organic produce in Budang (N. Nirola)

Kitchen gardens for improved well-being

Elbegzaya Batjargal and Tamana Zamir

In the high-altitude communities of Kyrgyzstan, the overwhelming majority of health problems affecting women and children are related to malnutrition. As shown by a project initiated in two districts of Kyrgyzstan in 2006, producing vegetables in kitchen gardens at the family farm level can prevent this problem by significantly improving nutritional and health status. They can also increase household incomes significantly – even in regions that have never grown vegetables before.

Izat Mashapova is tending her kitchen garden in Kara-Take village (A. Abazbekov kyzy)

In 2005, a health survey revealed that a large number of women and children in the Alai and Chong-Alai districts of Kyrgyzstan suffered from nutrient deficiencies. In the 20 villages surveyed, 78% of women and 48% of children suffered from anaemia, while 12% of children suffered from chronic malnutrition (1). Poverty in the area is widespread, with the socio-economic status of the population determined by the number of livestock, the availability of hay and the size of agricultural land.

Seeking solutions to health and poverty issues, a project was launched in 2006 to improve the nutritional status of women and children by introducing kitchen gardens in high-altitude communities (2,000 to 3,100 m) – gardens that would provide a variety of vegetables in areas that traditionally grew only potatoes. The project introduced vegetable cultivation, providing training and manuals, high-quality seeds and materials to build plastic tunnels to protect against low temperatures. Initially, three villages were chosen from each district. After three years, three new villages were added in the same districts, at which point support to the original three villages was cut back to only the provision of quality seeds.

The project has now established 310 kitchen gardens in 31 villages of Alai and Chong-Alai, 40% of which are cultivated without external support (2). The project has also introduced crop diversification and promotes crop rotation in order to prevent soil degradation, erosion, pest and disease problems, and phytotoxic effects.

The villages' kitchen gardens average only 0.01 ha, which is still enough room to grow tomatoes, peppers, beets, cabbage, carrots and garlic. In spite of this small size, 62% of the kitchen gardeners produce enough vegetables to sell part of their harvest at local markets. In 2011, the average kitchen gardener generated an

"*W*hen I was setting up my kitchen garden, villagers did not believe in the possibility of growing vegetables in our district, which is situated at an altitude of nearly 3,000 metres. Success didn't come right away. I participated in a series of trainings and exchange visits to improve my knowledge and skills in vegetable cultivation. Now I grow carrots, beets, peppers, tomatoes, garlic and greens."

Sharabidin Mashirapov, a member of the farmers' group in Jash-Tilek village

Lessons learned

- The stepwise process of growth associated with training, monitoring and gradually reduced external support is a key factor in the project's success.
- Introducing or promoting the production of vegetables at high altitudes contributes to improved access to nutrient-rich diets and improved health of local residents, and generates additional income. In addition, local production in remote areas increases availability and affordability of vegetables in local markets.

Family collecting vegetables from their kitchen garden (A. Abazbekov kyzy)

Taking care of the greenhouse located in Sopu-Korgon village (A. Abazbekov kyzy)

additional annual income of USD 280 from selling vegetables (2). In addition, 90% of the kitchen gardeners were able to preserve 30–50 kg of vegetables for their own consumption during winter, thus reducing dependence and expenditure on imported and processed food. Both the seasonal fresh vegetables and preserved vegetables contributed to the improved health status of project participants, with anaemia of mothers decreasing by 42 percentage points and of children by 39 percentage points (Figure 1) (3).

Production of vegetables in villages at high altitudes significantly reduces transaction costs caused by traders and middle men in the value chain of supply, and increases availability and affordability of vegetables in local markets.

Although minimal, the gardens face risks of unsustainability that the project is helping the farmers avoid by teaching them to rotate crops to avoid soil degradation, and about the importance of controlling the chemicals used for fertilizer and pest control, and the necessity of ensuring the availability of quality seeds.

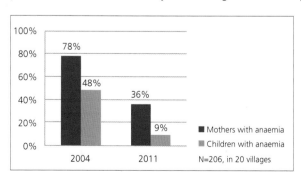

Figure 1: Anaemia of women and children decreased (3)

Organic farming improves income and diet

Gerhard Buttner and Cecilia Gianella

In Ayacucho, one of the most impoverished regions of Peru, the promotion of organic vegetables grown in greenhouses in combination with improved water and irrigation systems has contributed to an increase in income and a more diversified diet.

Preparing a new greenhouse for vegetable production in Ayacucho, Peru (G. Buttner)

In Ayacucho, where smallholder farmers make up most of the population, 68% live in poverty, 53% lack clean water and 79% lack basic sanitation. At the same time, 36% of the children are chronically malnourished – especially lacking fruit and vegetables in their diets. Per capita income ranges from USD 40 to 50 per month in what is, for many, mainly a subsistence economy.

The Centro de Desarrollo Agropecuario (CEDAP) runs an integrated rural development project in collaboration with the marginalized communities in the region, providing technical agricultural support and training, and combining Andean agricultural knowledge with modern agricultural practices. The present project, supported by UK Aid and Foundation Ensemble, supports families and farmers' associations in growing vegetables organically in greenhouses, by helping improve water and irrigation systems and enhance water security, which is endangered by climate change. By increasing food production for home consumption as well as for income generation activities, the project is increasing the resilience of small-scale farmers and their families and, hence, their livelihoods.

CEDAP uses family and community-level "contests" as an innovative tool to promote the use of local knowledge and integrate this with accessible new technology (Box). Called "Let's protect Mother Earth", the themes for the family contests include land planning, soil management and conservation, livestock management, agro-ecological technologies, sustainable water management, health, housekeeping and family education. In over 70% of the participating families, women take the lead in the competition, which contributes to their empowerment.

In 2012, one and a half years after the project had been launched, a household survey replicated a 2011 baseline study, interviewing 250 randomly selected house-

Family production contest (G. Buttner)

Lessons learned

- The implementation of greenhouses and irrigation has led to the creation of a growing local market in vegetables and milk derivatives.

- In addition to people directly involved in enhanced production, those who can now buy new products that enrich their diet in the local market also benefit. The local vegetable producers can sell their products at lower prices than the products that have been transported from the coast.

- Women play a vital role in this market, and also form the majority of the direct beneficiaries.

Carapo child enjoying a greenhouse carrot (G. Buttner)

holds in the 14 implementation communities and an additional 45 direct beneficiaries who formed part of the baseline sample. Instead of asking for exact income data, proxy variables were used to measure changes in income and consumption. This approach helped evaluate the changes in a context where it was difficult to capture credible data about the socio-economic situation of households.

- New assets: In 2011, only 20% of the households were able to acquire new assets. In 2012, this figure increased to 62% of the direct project beneficiaries, half of whom invested in productive assets in the form of agricultural tools.

- Surplus for sale: The surplus available for sale after subsistence needs had been met increased from 15% to 42%, confirming notably increased sales.

- Migration: Direct beneficiaries were slightly less prone to continue temporary migration than the general population.

- Consumption patterns: In 2011, only 82% of the direct beneficiaries had sufficient food to eat three meals a day. In 2012, the number increased to 95%, indicating a shift from below to above the region's average nutrition level.

- Dietary additions: In 2011, 93% of direct beneficiaries were able to diversify their basic diets with new vegetable varieties and thus had a more balanced diet and vitamin intake. This is higher than in the case of the overall population, of which only 80% were able to add new vegetable varieties.

How does the family contest work?

Each family creates a plan for the next 5 years on how it envisions improving production and developing land use and household organization. This is documented by drawing the present and the desired situation. The drawings are pinned to the wall in the famillies' homes together with a yearly month-by-month action plan. After six months, an evaluation committee – consisting of CEDAP, local municipal authorities and past winners – visits all participating homes to evaluate and score advances. Winners are invited to an award ceremony to receive their prizes of productive assets such as additional tools or irrigation material (co-sponsored by the project and the local municipality) and are invited to visit winners in other communities to further exchange their learning.

Sustainable mountain pastoralism: challenges and opportunities

Ajar herder transhumance to upland pastures in summer, Pakistan (H. Rueff)

Henri Rueff and Inam-ur-Rahim

A significant part of mountains and uplands is occupied by extensive pastoral systems allowing a large number of communities throughout the world to make a living. In the Hindu Kush Himalayan Mountains (HKH), for example, 60% of the land cover is rangeland (1). In Peru, 86% of the highlands are covered by pastures grazed by 70% of the country's livestock population (2).

Mountain pastoralists graze their animals according to season on a vertical landscape to produce high-quality livestock products sustainably. To do so, mountain pastoralists use marginal lands and practices of nomadism, transhumance and agro-pastoralism, the latter integrating fodder crop with grazing (3). These marginal lands would otherwise remain unproductive because their climate or topography makes them unsuitable for cultivation. Cultivated highlands may exist but often at the expense of environmental degradation.

Mountain pastoralists endure a number of stressors such as climate hazards and disasters (flash floods), and shrinking pastures due to competing forms of land uses. Mountains affect weather regimes, creating local climate variability over short distances. Mountain pastoralists receive little attention and investment, because they do not follow conventional production models and are poorly integrated into supply chains. As a result, they are often marginalized and do not have access to basic support services. Power relations between landowners and mobile pastoralists find the latter at a disadvantage. Policies tend to seek to "modernize" pastoralism through intensification rather than support these systems, which efficiently use marginal resources (4). Aid agencies still struggle to effectively support mountain pastoralism (5). As an example, land rehabilitation with monoculture afforestation in the past 20 years in the Pakistan HKH has obstructed transhumance routes for herders, forcing them to accelerate migration to upland pastures in summer. Early arrival at summer pastures fosters degradation because animals graze on grass sprouts. Afforestation programmes planting adapted tree fodder species could fulfil land rehabilitation needs while providing mobile pastoralists with fodder for their livestock when herds are moving (6).

Recent plots, first ploughed a year or two ago, are found higher on the steeper part of the slope without terracing.

Slope steepness increases with elevation accelerating soil loss

Older plots, first ploughed about 30 years ago, are found at the bottom of slopes in the most fertile areas

Crop encroachment onto pastures, Pakistan (H. Rueff)

Pastoralists provide regulating ecosystem services (climate regulation, flood and erosion control), provisioning services (food, water, genetic resources and fuel), cultural services (heritage and landscaping) and supporting services (nutrient cycling, habitat and primary production). A study conducted in the Naran Valley in the Pakistan HKH shows that high-altitude pasture management contributes more to climate change mitigation with a superior carbon store averaging 12.2 t C per ha as compared to cropping (7).

Building awareness about the services provided by mountain pastoralism could attract investments, improve development approaches and benefit herders and society. A workshop for landless mountain pastoralists, landowners, government officials and local academics held in Pakistan in 2012 resulted in the creation of a "pastoralism cell" within the Khyber Pakhtunkhwa Province ministry of agriculture. This body was charged with addressing mountain pastoralists' needs for support and voicing their interests through community-based organizations. This cell should also implement measures to secure transhumance routes by purchasing land for resting places. The University of Peshawar has also committed to promote pastoralism studies and to host pastoralist students by waiving tuition fees for their education (8).

Mountain pastoralists' economic contribution often goes unnoticed

Mountain pastoralism supports regional economies. For example, an estimate of pastoralists' production in Kyrgyzstan, a large part of which takes place in highlands, shows that the sector contributed 20.5% to the national income (9). The Ajar pastoralists, a community of about 7,400 landless households in the Khyber Pakhtunkhwa Province of Pakistan, market small ruminants worth USD 68 million per year (7). Mountain pastoralists also keep highly valuable livestock genetic resources with traits adapted to steep slopes, rugged terrain, poor-quality forage and resistant to diseases. These breeds are optimally suited to meet mountain pastoralists' needs for production, draught power and cash from sales. Their reproductive and productive performances relative to body weight are higher than those of exotic breeds (10, 11).

Lessons learned

- Keeping marginal mountain lands productive through pastoralism is a rational use of land especially when considering the pressing food security agenda in many developing countries.
- Governance of mountain rangelands needs to adopt an integrative approach, based on the awareness of the wide array of services provided by pastoralists beyond the supply of livestock products. This should be reflected by payment schemes and investments in mountain pastoral systems and sector-based approaches.
- Switzerland's payment scheme for mountain farmers has substantial returns on investment. Swiss alpine pastures managed by farmers "produce" the Swiss mountain landscape, attracting tourists globally and generating Switzerland's fourth-largest income source (12).

Vertical pasture landscape diversity in Tolök, Kyrgyzstan (H. Rueff)

Improvement of aquaculture practices in mountain farming

Nguyễn Hữu Nghĩa, Hoang Thu Thuy, Dang Thi Oanh, Tran Minh Hau

Protein from fish is an important part of the diet in many parts of the world – especially in Southeast Asia. In late 2012 and early 2013, an applied research project, supported by the UN Food and Agriculture Organization (FAO), focused on improving aquaculture practices on family farms and, in turn, made progress in improving the diet of local communities living in rural mountain areas.

Fish management and control (Hoang Thu Thuy)

Dak Lak Province in the Tay Nguyen Highlands of Viet Nam has over 500 dammed reservoirs and numerous rivers, making it a high-potential region for fisheries and aquaculture development. Between 2003 and 2012, as the area allocated to aquaculture increased steadily from 3,423 to nearly 7,000 ha, food quality improved, and production became more diversified.

In order to provide technology transfer to farmers, the FAO project included training courses and practical demonstrations on aquaculture. The project's fish fingerlings supply centre in the province built up a network for farmers and local extension staff who continued distributing fingerlings and provided technical advice after the project ended.

The project also surveyed the area's farming systems, technical skills of farmers, availability of fingerlings, agriculture by-products available for production of farm-made feed, and evaluated the status of fish ponds. The project used a SWOT analysis to identify demonstration sites for improved aquaculture, and then selected fish species to use in the demonstration sites according to whether they were:

- known and appreciated in the province, with high-quality, easily accessible seed,
- suitable for polyculture, i.e. for rearing of two or more non-competitive fish species in the same pond,
- able to be fed with farm-made feed from locally available agriculture by-products.

A meal with fish cultured by farmers (Hoang Thu Thuy)

Lessons learned

- Management is important. For example, ponds must be cleaned regularly, and also after rains, with lime for algae extermination and pH stabilization.
- Farm-made feed is sufficient to grow the fish, but adding manufactured feed – about 3–5% of the total feed volume – will increase fish growth.
- Project duration was too short. Farmers need more than six months to become familiar with aquaculture practices. It should have been extended to cover the rainy season when there is more water in the ponds, which is more productive for aquaculture.

Farmer's rich fish harvest in Lak district, Dak Lak province (Hoang Thu Thuy)

As a result, 40 farmers in the Tay Nguyen Highlands participated in the aquaculture training, with the project covering expenses for pond construction, fingerling distribution, production of small cages to catch the fish (*hapas*) and fishing equipment for 20 demonstration sites.

Fingerlings distributed to local households included grass carp, common carp and tilapia, all traditional species known to farmers. They can be grown easily and fed with grass, vegetables, cassava and bran, while the ponds can be fertilized using cattle manure with no or low external inputs. These fish species are also well adapted to local weather conditions. As transport in the area is difficult, the project initially supported farmers by transporting manufactured fish feed, but aquaculture experts also trained farmers in producing farm-made feed using locally available ingredients such as vegetables, grass, corn, cassava and rice bran.

The project found that after six months, fish reached an acceptable average 10.7 cm length and 160 g weight. Households harvested the bigger fish for home consumption, to present to relatives and neighbours or to sell in local markets. In general, the overall survival rate averaged 60% for fish species, but was 80–88% for tilapia and grass carp.

By the end of the project, farmers had learned the principles of aquaculture and the complete production cycle of main cultured species and were able to grow the fish by themselves using a combination of home-made and manufactured feed to increase fish growth. As a result, farming family diets contained more protein including cultured fish, shrimp and freshwater crab, and were further supplemented with food such as chicken, beef and vegetables bought in markets with money made from selling fish.

Organic farming as a climate change adaptation measure

Carla Marchant Santiago and Axel Borsdorf

The Río Las Piedras Basin in the Cordillera Central branch of the Colombian Andes has suffered from social conflict and also deals with the effects of climate variability. Today, thanks to a comprehensive strategy that encourages organic farming as a climate change adaptation approach, the basin has both reduced conflict and improved the quality of life for local farmers.

Río Las Piedras Basin landscape (A. Borsdorf, L. Ortega)

The Río Las Piedras Basin is located in the buffer and development zones of the Cinturón Andino Biosphere Reserve in Colombia's Cordillera Central. In 2001, local groups, supported by the regional government, launched a strategy to encourage organic farming in order to reduce poverty but also to promote peace in the area, which had been beset by guerrilla activities and conflicts over land use. As a result, peace returned to the area, enabling one of these groups – Asociación Campesina del Cauca (ASOCAMPO), an association of local farmers – to increase its focus on promoting sustainable farming and ecological agriculture. This helps farmers cope with climate change conditions, conserves the ecological integrity of the basin and, at the same time, promotes and fosters cooperation within the community. By 2011, 64 of the 97 farms in the basin had joined the association (1).

The area is highly vulnerable to climate change, with increased precipitation variability and extreme weather events occurring more often. Record high and low temperatures and increased precipitation have intensified erosion and crop losses due to pests and diseases.

ASOCAMPO has introduced several strategies for adapting to problems related to climate change since 2001, such as establishing forest patches to reduce flood impact, improve water retention and protect the raised bogs of the Andean wetlands, the *páramo*. It has also installed composting systems on its 64 member farms that are fed with dung, biomass and household waste. The resulting compost has replaced mineral fertilizers. Four farmers' schools in the basin produce 880 kg of compost every two to four months, which is more than the farmers need, so they can sell their surplus on the market. Thirty bags of 40 kg provide nearly USD 115 profit after accounting for labour costs (1,2).

Lessons learned

- The Río Las Piedras experience shows that organic farming has helped improve the livelihoods of farming families, even though their economic situation remains difficult.
- The project has also contributed to enhancing the social and human capital within the community and has contributed to peace in the area.
- The Río Las Piedras Basin is located in the development zone of a biosphere reserve (BR). Buffer and development zones of a BR are ideal places to implement such initiatives. The Río Las Piedras initiative could be replicated in other communities living in rural mountain areas with similar problems.

Farmers in the ASOCAMPO assembly (A. Borsdorf, L. Ortega)

In order to reduce erosion due to livestock trampling, ASOCAMPO has led subdividing of pasture with electric fencing and introduced rotational grazing. Trees have been planted to prevent wind erosion, steep slopes terraced using organic material such as bamboo and acacia wood, and tree nurseries established to grow indigenous trees. Bioengineering techniques introduced in road maintenance help prevent erosion and landslides and ensure year-round use, and market gardening and fruit production now use greenhouses and drip irrigation. The sustainable farming strategy also included education for climate change adaptation, and different policies for encouraging sustainable agriculture based on local knowledge (3).

In 2011, ten years after introducing organic farming to the basin, a survey found considerable improvements in farmers' livelihoods, but their financial situation remained difficult, as farm production – mainly milk and cheese – was mostly used for household consumption. Most of the families would invest in the extension of their farm if they had the means to purchase more land, livestock or machinery. Poor accessibility remains a challenge, as roads are not paved, which raises local transport costs. Overall, the farmers' situation remains challenging, despite the merits of organic farming (1).

Composting is essential for organic agriculture (A. Borsdorf, L. Ortega)

Biosphere reserves as an instrument for sustainable rural development

Biosphere reserves are the central instrument of the UNESCO Man and the Biosphere Programme, which started in 1976 to promote sustainable regional development. Biosphere reserves are large and representative portions of natural and cultural landscapes, which should be secured for the longer term. They are meant to present a refuge for genetic resources and ecosystems on the one hand and a model for sustainable use of land, education, research and recreation on the other.

Mountain products and market development

"Marcha Eco Solidaria" – female farmers in La Paz, Bolivia, march for fair trade

Farmers in Lao PDR process river weed for the market (Ch. Flint)

Susanne Wymann von Dach

Mountain products and market development

Smallholder farmers in mountain areas, who are weakly integrated in commodity markets and hardly able to compete with large-scale producers from lowlands, now have an entry point to capitalize on emerging markets for nutritious, healthy and organic products. These emerging markets offer windows of opportunity for developing pro-poor sustainable value chains, thanks to labelling and formal certification schemes that guarantee the value added of mountain products and help bring premium prices.

When it comes to market participation, smallholder mountain farmers are hampered by low, dispersed and unreliable production levels, remoteness, lack of processing technology and knowledge, and difficult access to market information, as well as inadequate negotiation and management skills. It remains difficult for them to make sufficient money to meet their basic needs, invest in their farm infrastructure and fulfil personal aspirations.

In order to adapt to their challenging environment, mountain farmers have developed highly diverse farming systems by integrating crop production with livestock, forestry and fishery, which may now turn their seeming disadvantage into a comparative advantage. They have respected cultural diversity, resisted homogenization of their products, domesticated crops and livestock, created and conserved agro-biodiversity and thereby developed in-depth local knowledge about usable wild species (Table 1).

They have also managed these integrated farming systems with low input of chemical fertilizers and pesticides – all of which adds up to the potential for producing attractive, healthy and organic food for new markets. Consumers, including mountain tourists, and the private sector are re-discovering the highly nutritious and medicinal value of indigenous, underutilized and wild species. They appreciate the qualities of organically grown or speciality products, and are willing to pay premium prices. At the same time, urbanization in some mountain areas offers markets for locally grown products.

Now, as a next step, it is necessary to develop value chains that enable family farmers and particularly poor households to participate in and benefit from these emerging markets. Such value chains need to be developed jointly by representatives from all stakeholder groups and based on a sound analysis of the mountain-specific challenges, natural resources and market potential as well as the farmers' socio-economic capacities and the relations among the value chain actors (1). Moreover, the development of a new value chain must not jeopardize the farmers' own food security and sustainable production systems (2, 3).

Mountain farmers would undoubtedly benefit from capacity building aimed at developing technical and managerial skills, promoted by both the public and private sectors. Collective action is key to overcome shortcomings of unreliable and low productions, and enhances the producers' negotiation power in the value chain. More direct links between producers, sellers and consumers will benefit farmers but will also reduce their vulnerability to exploitive practices of traders and middlemen. Moving ahead, appropriate technologies and infrastructure, such as decentralized and renewable energy supply, will be required to establish or enhance processing activities in mountain areas and, in turn, provide off-farm jobs. Producers and processors need to communicate the quality, uniqueness and origin of their products to the consumers in order to obtain higher prices that will cover the high labour input needed for maintaining ecosystem services provided by mountain areas (4). While the process of labelling and certifying products (Box) guarantees quality and traceability of mountain products, it also entails considerable communication and administrative efforts that can be beyond the ability of marginalized smallholder farmers – who then risk being excluded from promising markets or unable to take advantage of them. Therefore administrative requirements of formal certification schemes should be kept to a minimum without threatening the credibility of the scheme. Often labelling is sufficient for small production volumes that are meant for local and regional markets, while produce for national and global markets can garner a better cost–benefit ratio when formally certified.

Establishing niche markets under the prevailing liberal market regime in many countries requires enabling policies that acknowledge the added value of mountain products as a means to improve mountain livelihoods and regional development and at the same time compensate the higher labour input for maintaining critical ecosystem services.

Sapsago: a branded mountain product for 550 years

In the fifteenth century, a blue fenugreek flavoured hard cheese known as sapsago was the biggest-selling product of Glarus, a Swiss mountain valley. Its main market at that time was the city of Zurich. In 1463, the people of Glarus established regulations for sapsago production, establishing it as a brand. The brand allowed them to distinguish their product from the cheeses of competitors, and to guarantee its quality and obtain a premium price. The branding was such a success that by the seventeenth century, it was necessary to limit export in order to secure sufficient supply for Glarus and to reduce speculation. Despite marketing crises over the centuries, sapsago remains an important export product, sold in more than 50 countries (9, 10).

Promoting the branded speciality cheese sapsago to city dwellers (GESA)

Agro-biodiversity	Survey region	Record
4,000 varieties of native potatoes	Andean highlands of Peru, Bolivia, Ecuador	International Potato Centre (5)
1,299 species of medicinal plants	Gaoligonshan Nature Reserve, China	HKH* conservation portal, species data set of ICIMOD (6)
600–700 non-timber forest products (NTFP) (plant species only)	100 upland communities in Luang Prabang and Xien Khouang Province, Lao PDR	NTFP database of TABI** (7)
131 different livestock breeds	Turkey (nationwide)	FAO Domestic animal diversity information system (8)

Table 1: Selected examples illustrating the high agro-biodiversity in different mountain regions
*HKH – Hindu Kush Himalayan; **TABI – The Agrobiodiversity Initiative

Certification frameworks for mountain products

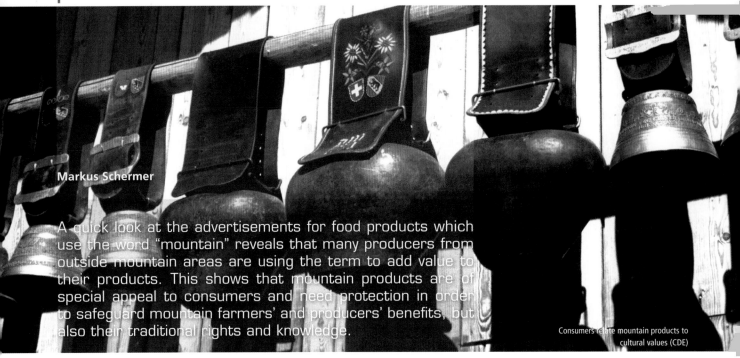

Markus Schermer

A quick look at the advertisements for food products which use the word "mountain" reveals that many producers from outside mountain areas are using the term to add value to their products. This shows that mountain products are of special appeal to consumers and need protection in order to safeguard mountain farmers' and producers' benefits, but also their traditional rights and knowledge.

Consumers relate mountain products to cultural values (CDE)

What are mountain products? The answer seems to be clear: products coming from mountain areas. But do all the ingredients have to come from mountain areas or only the major ones? Do all the processing stages have to take place in mountain areas or only the ones adding the most value? And how do consumers perceive mountain products?

The research project EuroMARC (1) conducted the first study on consumer perception in six European countries. European consumers (in mountain and non-mountain areas, cities and rural areas) have a common understanding of what mountain products are (Figure 1). They associate mountain products with health or purity and with special products. In all six countries and for all product categories the following criteria were most important for the consumers: the taste and origin of mountain quality food products. However, mountain quality products stand for more than just food products and are closely associated with the culture of mountain people. The products represent the combination of many attributes: food, mountain areas, nature, local production, nostalgia. European consumers currently cannot distinguish between products that are correctly labelled as products coming from mountains and those that are taking advantage of people's positive view of mountains for promotional purposes. Thus, a regulation on the use of the word "mountain" seems necessary.

There are different ways to protect mountain products against falsification. The issue of mountain products has already been addressed by the European Union, which reserved an optional quality term (Box). This regulation is the result of a long lobbying process by Euromontana, which has issued a Charter on Mountain Products (2). However, for the time being, the precise definition and delineation of mountain areas remains largely a national responsibility and thus differs from country to country.

Many retailers regard the denomination "mountain" as too general and prefer a more particular designation of provenance. They also express the concern that

Lessons learned

- The optional use of the reserved term "mountain product" may assist farmers in mountain regions in better communicating the special characteristics to the consumer, without much administrative burden, in contrast to labels. Thus this may be an approach that is more feasible for small-scale producers. However, the effectiveness needs to be ultimately proven by practical application.

rigid regulations, which are necessary to meet consumers' expectations, may exclude larger producers, while small producers might have little interest in additional certification procedures and costs (1). Such certification schemes exist already, such as Protected Denomination of Origin (PDO), Protected Geographical Indication (PGI) or Traditional Speciality Guaranteed (TSG). They all offer the possibility of communicating the qualities of products better to consumers and bringing more added value to mountain areas. However, of the 1,076 currently registered PDO/PGIs in the EU, only 171 can be considered "mountain" products while 196 are "partially mountain". Very often the processing occurs in or partly in mountain areas and raw materials come from other areas (3).

At the same time, a number of European countries are working towards policy frameworks that provide clear guidance for but also the protection of "mountain products". Italy and France have national laws defining "mountain products". Outside the EU, Switzerland has definitions for "mountain products" and "alpine pasture products". The main criteria defined by these laws concern the areas of origin of raw materials, and the processing and packaging of products. Recently, a working group (the Mountain Farming platform) of the Alpine Convention (www.alpconv.org) has become engaged in defining the "alp" and "mountain" terms as reserved terms within the framework of the Convention.

EUROPEAN UNION DEFINITIONS

"Mountain areas" (4)

"Mountain areas shall be those characterized by a considerable limitation of the possibilities for using the land and an appreciable increase in the cost of working it due to:

- the existence, because of altitude, of very difficult climatic conditions, the effect of which is a shorter growing season;
- a lower altitude, to the presence over the greater part of the area in question of slopes too steep for the use of machinery or requiring the use of very expensive special equipment, or
- a combination of these two factors, where the handicap resulting from each taken separately is less acute but the combination of the two gives rise to an equivalent handicap."

"Mountain product" (5)

"1. The term [mountain product] ... shall only be used to describe products intended for human consumption ... in respect of which:
   (a) both the raw materials and the feedstuffs for farm animals come essentially from mountain areas;
   (b) in the case of processed products, the processing also takes place in mountain areas."

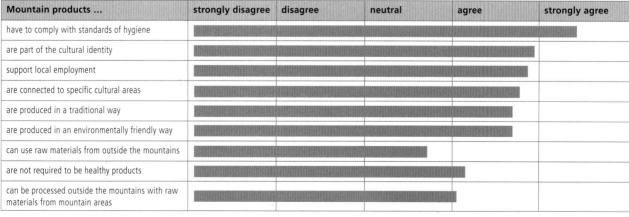

Figure 1: European consumers' expectations of the quality of mountain products (6)

Agribusiness development through cooperation

Dyutiman Choudhary and Mahendra Singh Kunwar

Due to poor market access, low production output and their lack of information, capital and services, farmers in Uttarakhand, India, traditionally received low prices for the Malta oranges they produced. However, a farmers' federation helped increase production while a cooperation based on farmers' self-help groups has enabled the processing and marketing of these fruits, increasing the farmers' incomes threefold.

Ripe Malta oranges (M. Singh Kunwar)

Farmers producing Malta oranges in Uttarakhand traditionally dealt with a value chain that was disorganized and non-competitive, and ended up selling their products to traders at low prices. Even a state government programme to purchase the fruits at a minimum price failed as farmers had to wait for payments up to six months after selling their produce.

Two kinds of support groups were formed that have strengthened farmers' collaboration and improved their positions in the value chain (Figure 1).

- Farmer interest groups (FIGs) were formed individually but joined forces as a federation. Now, the federation collects the farmers' products at production sites, and the accumulated volume increases the farmers' bargaining power.
- Self-help groups (SHGs) were formed to support local women, linking them to banks for loans to finance processing activities. In total, 27 such groups were established and, with the support of the Himalayan Action Research Centre (HARC), were united to form a cooperative. One member of each SHG participates in the cooperative, which now manages a Common Facility Centre (CFC), which processes Malta oranges and other local agricultural produce (1, 2).

The cooperative creates employment for SHG members in processing five types of products from Malta oranges, such as juice and marmalade, as well as in packaging the products and labelling them with a brand. The cooperative sells the

> **Malta oranges (*Citrus sinensis*)**
>
> Malta oranges are grown abundantly in the Indian state of Uttarakhand. The fruit is cultivated at altitudes between 1,000 and 2,000 m. It was introduced and promoted by the state's horticulture and watershed departments. However, when harvested, the Malta oranges face intense competition from citrus fruits produced in Western India. Because of their sour taste and thick skin, Malta oranges are often not preferred by consumers for use as fresh fruits but they are competitive for other uses.

value-added products under the brand name "Switch On" to a large number of pilgrims who travel through the region, and in the local and regional markets. It also charges a fee of 3% of the SHGs' production costs to cover its operational expenses and ensure its own sustainability. The women's groups deposit the profits from their activities into their revolving funds.

The combined efforts of the federation and cooperative, with the support of experts and consultants, have assisted members in improving the quality and quantity of their produce by organizing training and demonstration programmes on production, grading, processing and packaging. These interventions have also increased the confidence and capabilities of small farmers – with tangible results. For example, in 2011, 539 farmers sold 150 tonnes of fruit to local markets for Rs 8–10 per kg (USD 0.17–0.21) compared with Rs 1–2 per kg before the intervention. The farmers sold also low grade Malta oranges to the cooperative whose overall turnover increased significantly (Table 1). As a result of the interventions, farmers' income from Malta oranges increased threefold. The Government of Uttarakhand recognized the potential of the intervention and increased minimum support prices – a measure taken to ensure minimum prices to farmers – from Rs 5.25 to Rs 6 per kg (1,2).

Based on its experience with Malta oranges, the cooperative has already diversified its production to include other local products. Now, with its increased revenues and profits, the cooperative is expected to expand its membership, enabling even more mountain farmers to receive higher prices for their production.

Lessons learned

- Increasing farmers' bargaining power, capacities and access to information can lead to retention of a higher value share in mountain regions by enhancing their "terms of participation" in the value chain.
- Markets are a key factor in sustaining the cooperative and enterprises that process the fruits. In the long run, cooperatives need to maintain and upgrade the quality of their products continuously and further expand their market networks. This indicates that capacity building on various aspects of institutional management, compliance with laws and policies, product development and exposure visits has to be continued.
- Providing a business platform, local employment opportunities and increased income can substantially improve the livelihoods of mountain women and men.

This model for agribusiness development can be replicated in other mountain regions with pro-poor, inclusive and mountain-sensitive policies.

Products	2009		2011	
	Volume	Revenue (Rs in ,000)	Volume	Revenue (Rs in ,000)
Raw Malta orange peeling	50 t	90	200 t	200
Juice-making	50 t	20	200 t	33
Marmalade	0.3 t	30	1.5 t	150
Peel powder	0.2 t	10	0.5 t	50
Squash	5,000 l	75	20,000 l	300

Table 1: Development of turnover of the cooperative
Source: Cooperative record (USD 1= Rs 47 in 2009 and Rs 48 in 2011)

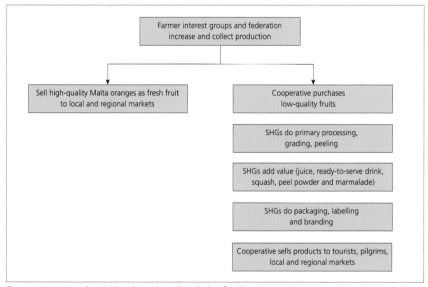

Figure 1: Actors and activities along the value chain of Malta oranges

Farmers pack Malta oranges for processing (M. Singh Kunwar)

Adding value to traditional mountain crops

A field of quinoa in the Southern Bolivian Altiplano (S. Padulosi)

Alessandra Giuliani and Stefano Padulosi

Across the centuries, family farmers in the high Andes have selected and adapted varieties of resilient indigenous grains that have reduced their vulnerability to environmental risks and provided valuable nutrition. Today, a fast-growing national and international market for these grains has brought with it the challenge of combining production for markets without threatening people's food security and the environment.

Temperatures in the high Andes can range from -18 °C to 27 °C during the day with overnight frost for more than 200 days a year. At the same time, rainfall averages less than 250 mm a year and soils often have low fertility. Due to these harsh conditions, only few crops can be grown such as quinoa (*Chenopodium quinoa*), cañahua (*Chenopodium pallidicaule*) and amaranth (*Amaranthus caudatus*). These are species that adapt well to drought, floods and frosts – conditions that are likely to increase under climate change. Quinoa, cañahua and amaranth are also categorized and defined as "neglected and underutilized species" (NUS).

For centuries, family farmers in Bolivia, as well as in Peru and Ecuador, have relied on these grains that still play an important role in Andean nutritional security. They have a comparative advantage over other staple crops in terms of resilience and having high levels of protein and micronutrients (1) (Table 1). In recent years, consumers worldwide have paid increased attention to these healthier, nutritional and traditional food products, and they have become an important source of income (2). In some villages, quinoa now accounts for more than 80% of family farms' agriculture income.

Research to boost the marketing of Andean grains has been carried out in the framework of a global project supported by the International Fund for Agricultural Development (IFAD), implemented by Bioversity International in collaboration with research agencies in Bolivia and Peru (3). In Bolivia and Peru more than 1,170 families participated directly in the implementation of the initiative, which demonstrated the value of NUS and their current uses. Today, many new products are

"*W*ith the use of the community quinoa processing equipment, I have a surplus of quinoa to sell. With this additional marginal income I can now send my children to study in a nearby bigger village."

Quinoa family farmer in Colcha, Bolivia

Component	Quinoa	Cañahua	Amaranth	Wheat	Rice	Maize
Protein	12.6–18.4	11.6–19.5	10.2–18.3	8.6	6.6	8.7
Fat	4.2–8.7	1.7–8.9	4.5–12.8	1.5	0.4	3.9
Carbohydrate	54.3–73.0	53.4–72.7	66.5	73.7	80.4	75.7
Fibre	3.5–8.0	4.1–18.5	6.6	3.3	0.7	2.4
Ash	2.1–4.7	3.1–22.1	2.1	1.7	0.8	1.5

Table 1: Nutritional composition of Andean grains (amaranth, cañahua and quinoa) vis-a-vis wheat, rice and maize (g/100 g). Source: (2)

Women in Bolivia are introduced to the use of a quinoa saponin removal machine (S. Padulosi)

Lessons learned

- Andean grains as well as many other NUS have great potential for contributing to improved food security and nutrition, and for increasing income, particularly for communities in marginal mountainous regions, as family farmers can be involved in all steps of the value chain.
- More attention by policy-makers is needed to support the development of NUS value chains through an enabling policy environment.
- In order to avoid negative repercussions on local production systems and support the livelihoods of local populations, advocated incentives should be accompanied by sustainable cultivation practices, measures to foster partnerships among value-chain actors and interventions to build capacities of isolated communities to better seize market opportunities.

derived from Andean grains such as flakes, cereals, juices, soups, biscuits, sweets, pasta, marmalades, cooking oils, cosmetics and pharmaceuticals (soaps, insect repellent), but more could be developed. In spite of the booming foreign markets, in particular for quinoa, the agribusiness of Andean grains is still very limited.

Family farmers are at a great disadvantage compared with large-scale commercial farmers. Not only do they often face significant transaction costs in marketing Andean NUS and lack access to processing facilities and skills in good management, they also have restricted access to capital, education and market information about consumers' needs and marketing institutions. Often the markets are poorly defined, value chains disorganized and demand is weak because the farmers' products are less well known and limited owing to difficulties in logistics, traceability and communication (3).

In Bolivia, the project helped develop a micro-processing technology for removing the bitter-tasting saponin coating from the quinoa grains, which has greatly reduced what was a laborious and time-consuming process. In addition to increasing quinoa consumption and income generation, farmers may also have the potential to market the saponin as a by-product (Box). The value chain has been restructured in order to improve efficiency and farmers' accessibility. To that end, the project promotes multi-stakeholder collaborative platforms that look at ways to channel the benefits to the family farmers. The platforms link producer associations, research institutes, development agencies, civil society, the business sector and policy-makers, helping build trust among value-chain actors, while advocating for greater policy support (4).

New quinoa products (J.L. Soto Mendizábal)

> Each year, quinoa producers throw away saponin (the bitter coating of the quinoa grains) with an estimated value of up to USD 5 million. Saponin can be used as a detergent or as an antiseptic. The problem is that there is no access to speciality markets for this product. With appropriate technology, information and improved market channels, 15,000 poor family farmers organized into cooperatives could bolster their income by producing saponin.

Spinning a fine yarn

Barbara Rischkowsky and Liba Brent

Angora goat production and mohair marketing are vital for rural households in northern Tajikistan. Yet, poor access to global markets and inadequate services threaten the long-term viability of the sector, with farmers losing hundreds of thousands of dollars annually in potential revenue. An International Center for Agricultural Research in the Dry Areas (ICARDA) project that has trained women spinners in processing kid mohair into luxury yarns for export, and farmers in improving goat breeding and fibre quality has greatly increased local income.

Angora breeding bucks in northern Tajikistan (L. Brent)

Rural communities in remote areas of Tajikistan are cut off from lucrative global markets for their mohair production. To produce the raw material for quality yarn, for example, farmers need well-fed, healthy animals that produce superior fleeces. Spinners and knitters need to know what luxury products that bring high prices look like and how to make them. They also need to know how to advertise, fill orders and ship what they have made. An IFAD-ICARDA project that has supported local people in meeting all of these needs has increased the ability of the goat and sheep sector to improve livelihoods in marginal and mountainous regions of Kyrgyzstan and Iran as well as Tajikistan.

The project has worked longest in northern Tajikistan, where people rely on Angora goats and mohair marketing for their livelihoods. Mohair provides an earning opportunity for smallholder farmers, and for the rural women who spin mohair into yarn, their most important source of income. For historical reasons, Russia still buys over 70% of Tajikistan's mohair produced by adult goats. But Russia has no processing capacity for kid mohair used for luxury yarns and textiles that are highly prized on the world market. The isolation of rural Tajik women effectively cuts them off from these markets. And unlike farmers in South Africa, Australia and Argentina, they are not supported by breeding and extension services and have no marketing infrastructure. Not only is a huge potential untapped, without support, the mohair sector might collapse – with dire consequences for thousands of families whose livelihoods depend on it.

The project started in 2006 with a programme to add value along the entire market chain. At the start of the chain, livestock scientists worked with farmers to create breeding goats that produce finer mohair. Farmers learned how to manage their flocks, how to improve feeding regimes and keep the animals in good con-

Women's group working on mohair dehairing (L. Brent)

Lessons learned

- The experience in Tajikistan showed the need to address all components of the production system, from animal breeding, livestock husbandry to processing and marketing. This integrated approach benefits men and women.
- By learning how to meet the standards of quality markets, mountain communities can capitalize on a growing global demand for natural, handmade products, including mohair, cashmere or wool, and develop profitable export chains.
- Cottage industries based on mohair, cashmere and wool oriented to global markets can boost incomes in remote communities provided they are supported with a longer-term perspective.

Spinning mohair at home using solar-powered spinning machines (L. Brent)

dition. The project collaborated with local and international breeding experts on creating breeding nuclei on selected farms, which then sold or lent the animals to other farmers. The project also tests mohair samples and evaluates mohair based on international standards, and links the farmers with local spinners' groups that are willing to pay higher prices for quality mohair.

Professional knitters in the USA tested samples of the yarn and provided feedback to the Tajik women. Although quality yarn takes longer to produce than the yarn the women produced before, it brings much higher prices. To generate even further profit, women were taught how to knit items such as shawls and sweaters that sell well in global markets – an eye-opener for the women, as they had never seen high-quality yarn or luxury goods before. Producing for the Russian market, they made USD 4 per kg from spinning mohair into yarn, but USD 52 when producing fine yarn for the USA market. The Tajik women now train women from other areas in Tajikistan and from Iran, and are receiving further training in how to set up businesses to expand their nascent cottage industry. This will involve linking women's groups with buyers in the USA and Europe and setting up ordering and shipping systems.

The project trained three groups of 55 rural women to produce high-priced, high-quality kid mohair yarns for yarn shops in the USA and Europe, and six groups of 29 women in knitting and weaving. Already skilled in spinning coarse mohair with spindles and knitting intricately patterned traditional socks and mittens, the women were introduced to spinning wheels and electric spinning machines, and now knit items that appeal to luxury export markets. Women also learned how to choose good fibre, and clean and prepare it for spinning. They can now spin raw mohair into quality yarn that meets international standards.

Community-based beekeeping for better livelihoods

Uma Partap and Min B. Gurung

In the Hindu Kush Himalayan (HKH) region, home to a rich diversity of indigenous honeybees, beekeeping is an integral part of the mountain farming systems. Organizing beekeepers, building their capacities in movable-frame hive beekeeping, and facilitating market linkages have led to a significant increase in their honey production and had a positive impact on their livelihoods.

Practical training in beekeeping (M. B. Gurung)

The indigenous honeybees of HKH help increase crop productivity, conserve biodiversity and can contribute to community well-being (Box). Beekeeping's high-value, organic, natural and ecologically sound products, such as honey and beeswax, provide important sources of cash income, but also have cultural value to local societies and fit in well with mountain specificities.

The Alital area of Dadeldhura District in Far West Nepal has a rich tradition of beekeeping thanks to the indigenous honeybee *Apis cerana*, which was traditionally managed by individual beekeepers in fixed-comb log or wall hives. However, honey was harvested from these types of hives by squeezing the combs, which not only resulted in low yield and poor-quality honey, it could kill the brood and adult bees, leading to a decline in colony strength, and keeping honey harvested in Alital from entering the mainstream market.

In 2000, the International Centre for Integrated Mountain Development (ICIMOD) initiated community-based interventions to address key issues related to honey production, quality, marketing and income generation. The project organized village-based training to build the capacity of local beekeepers to act as trainers and provide follow-up support. This training, plus the beehives and related equipment were offered to local people at no cost, in order to ensure the involvement of traditionally low-status castes, women and economically disadvantaged groups that had difficulty attending training outside their villages. Other key activities included conserving and planting bee flora, enhancing access to savings and credit and training in colony management, as well as developing market linkages focused on honey collection, processing and packaging as well as product promotion for direct selling to retailers and consumers.

The project also introduced movable-frame hives, from which higher-quality honey can be harvested, and trained two beekeepers in hive-making, who then established hive carpentry workshops, selling their hives for around USD 20 per hive. Beekeepers initially organized themselves through informal community groups. In 2005, these groups were supported in developing the Alital Multipurpose Cooperative Limited, which now has 117 shareholders from 12 villages and has been strengthened to provide services such as training and marketing support to beekeepers. Beekeepers also developed market linkages by participating in exhibitions and honey festivals. The Alital Chiuri Honey brand, which promotes its better processing and quality assurance, is now sold in Kathmandu, the country's capital.

Apis cerana beekeeping has now become an important income-generating activity for the people of Alital (1, 2) (Table 1). The number of farmers who have adopted movable-frame hives, the number of colonies in movable-frame hives and per household, as well as the annual honey production have increased substantially. The beekeepers earn now up to USD 4.5 per kg of honey, more than double the profit from 2001. This additional income is helping them buy clothes, oil, salt and, more importantly, pay school fees and buy books for their children (3).

Currently, Alital has the only beekeeping resource centre that provides bees, beehives and training services to farmers and organizations working in the Far Western region of Nepal.

Impact	2000	2012
Number of beekeepers adopting movable-frame hives	1	117 out of approx. 490 households (97 men and 20 women)
Beekeepers' organization	None – unorganized	Village-based groups forming the Alital Multipurpose Cooperative
Total number of bee colonies in movable-frame hives (number per household)	6	>1,000 (5–42)
Annual honey production	100 kg	More than 2,500 kg (21 kg per beekeeper)
Contribution to household income (sale of honey and sale of bee colonies)	Negligible	35–50% (on average USD 420 per household)
Hive-making workshop	None	2 carpenters (each earning USD 1,000 per year)

Table 1: Tangible results of project interventions. Sources: (1,3)

Improving the use of honeybees for apple pollination

An ICIMOD project in Himachal Pradesh promoted the use of honeybees for managing apple pollination (4).

A well-organized system has been established for hiring and renting honeybee colonies for apple pollination, where the Department of Horticulture assesses the demand for honeybee colonies for pollination and makes supply arrangements with private beekeepers. Both *Apis cerana* and *Apis mellifera* are being used. The current rate for renting an *Apis cerana* or *Apis mellifera* colony for pollination is between USD 13 and 17 per colony for the flowering period of apples. The income of apple farmers has increased as their goods are able to fetch higher prices thanks to the boost in crop productivity and improvements in fruit quality as a result of bee pollination. Thus, the use of honeybees for pollination of cash crops has proved to be of great benefit to both the beekeepers and the farmers. The large-scale use of honeybees for apple pollination has led to a new vocation: independent pollination entrepreneurs. They complement the governmental services.

Lessons learned

- The initial success shows that a community-based approach for improving local beekeeping by strengthening the capacity of local institutions can upgrade the value chains of high-value products and lead to increased cash income and improved well-being of family farmers.
- The micro-level model of development using local resources can be replicated in similar villages in the Hindu Kush Himalayan region.
- Increased private-sector engagement could lead to further improvements in the quality of local honey and could enhance its branding as an organic mountain product, which could improve access to national and international markets.

View of a village in the project site (D. Tandukar)

Diversification of mountain livelihoods

Tajik woman is weaving on their summer pasture in the Murghab District (K. Wolfgramm)

Landscape around Alto Beni, Bolivia (J. Jacobi)

Diversification of mountain livelihoods

Thomas Kohler and Kata Wagner

Family farmers' livelihood strategies are informed by the geography, history and culture of their environment, and by the political and economic frameworks of their countries. Mountain livelihood strategies have always required specific levels of resourcefulness, adaptation and diversification of income opportunities.

Mountain family farmers are exposed to the whims of weather, crop and animal diseases, changes in agricultural input and commodity prices, and shifts in policy and regulatory frameworks. They thus face the same risks as their counterparts in lowlands. But mountain farmers are often additionally burdened with shorter vegetation periods, steeper slopes and more shallow soils, a higher risk of ice, snow and hail, and the occurrence of landslides and avalanches. In response, mountain farmers have adopted risk-averting and risk-spreading strategies that have led to complex and diversified farming systems, using different resources – cropland, pastures and forests – at different altitudes and at different times of the year.

In many regions, farming forms the backbone of mountain farmers' livelihoods. Farmers produce for home consumption but also for the market as a source of income. Even in the most remote places, farmers need cash for health and education expenses, and for purchasing basic items they cannot produce themselves. Mountain farmers seize opportunities for income diversification, both on-farm and off-farm, to stabilize and increase their income and to enhance their livelihoods.

Diversification is often not a choice, but a necessity for farming households in mountain areas that are driven by population pressure, land shortage, natural disasters, hunger and poverty. Globally, only 22% of mountain areas are suitable for crop production. Looking specifically at the mountain areas in developing and transition countries, the percentage of cropland falls even lower, to a mere 7%.

Globally, population density on grazing land at all elevations up to 3,500 m has reached or surpassed the critical point of 25 persons per km². And about half of the 300 million people who are food insecure in the world's mountain regions suffer from chronic hunger (1).

Looking at specific aspects of diversification strategies, farmers in the mountains of Badakshan Province, Afghanistan, for example, have a wide array of income-generating activities typical for many mountain communities. The data shown in Figure 1 are based on a survey carried out among 26 remote and 22 non-remote villages, and document the decisive role of remoteness, especially its negative effects on non-farm income opportunities (2). Remote villages depend to a larger extent on farm income, but for both groups, farm incomes are lower than non-farm incomes, which include salaried incomes, self-employment and remittances.

Coca leave and flower offering to the *Pachamama* (Mother Earth), Pitumarca, Peru (S.-L. Mathez-Stiefel)

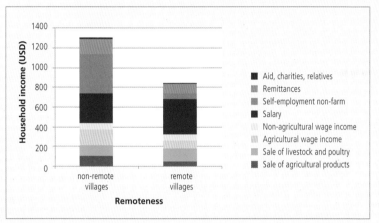

Figure 1: Average annual household incomes (medians) by source of income and remoteness, Badakshan, Afghanistan. Source: (2)

(N = 490 households in 22 non-remote villages, 614 households in 26 remote villages)

In contrast, farmers at Alto Beni, Bolivia, largely rely on farming for their livelihoods. This is possible because they mainly grow cocoa, a cash crop and high-value niche product (Figure 2). Income data gathered in a survey of 30 organic and 22 non-organic farmers reflect the premium price paid to producers of organic cocoa, which results in 40% higher incomes compared with non-organic production (3, 4). The cocoa produced is processed to chocolate to serve the Bolivian market; the chocolate from organic cocoa is also exported.

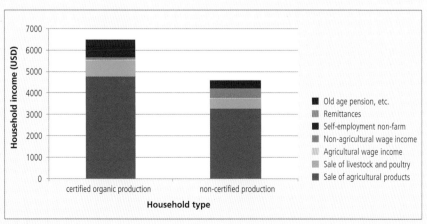

Figure 2: Average annual household incomes (medians) by source of income and mode of production, Alto Beni, Bolivia. Sources: (3, 4)

(N = 30 households with certified organic and 22 households with non-certified production (cocoa))

Village with gardens, Badakshan, Afghanistan (A. Pain)

In the uplands of Viengkhang District, Lao PDR, household strategies, farming systems and incomes show great variation within the same area (Figure 3). A survey of the livelihoods of farmers in the district identified four household types (5): (a) shifting cultivators, who practise the traditional mode of farming and grow rice for subsistence, (b) households that have moved from shifting cultivation to rotational rice cropping and that also keep large livestock such as cattle and buffalos, (c) diversified producers who have added plantations to their portfolios, mainly rubber and teak for export to China, Thailand and Vietnam, but who retain rice production, and (d) households that focus almost exclusively on rubber and teak plantations, keep livestock, but have given up rice cultivation, a key element of farming and culture in the region. Plantation farming, which appeared in the mid-1990s in the region, increases local incomes and income disparities very significantly, as shown by the income gap between more traditional farm types and farmers with plantations. Focusing on plantations also means less diversity of production and increased dependency on global commodity prices. Questions of sustainability also arise, relating to the effects of plantations on soil erosion, water quantity and quality, biodiversity and household food security.

In addition, remittances from migrated family members make a significant contribution to income in many mountain regions, such as Central America, the Andes and the Hindu Kush Himalaya. Migration also has proven to be a means for reducing dependency on local resources and acquiring new skills. As it is often male family members who outmigrate, women are left as managers of family farms. Tourism offers significant employment and income opportunities in mountain areas, especially in high-income countries but increasingly in the developing world. Mountains' clean air, diverse landscapes, rich biodiversity and unique cultures

Lao farmer harvests cassava, a cash crop (U. Wiesmann)

attract 10–15% of the global tourism market (6). Payment for ecosystem services is also an important element of family farm incomes in many high-income countries such as Switzerland, Japan, Norway and Iceland.

Opportunities to diversify and enhance mountain family farming livelihoods are manifold. However, taking further advantage of these opportunities will require an enabling policy framework in support of sustainable mountain farming – a framework that should include facilitation of payments for key ecosystem services, investment in capacity development for the empowerment of rural populations, in particular of rural women, and development of a network of decentralized small towns to provide markets, employment and vital services to rural mountain communities.

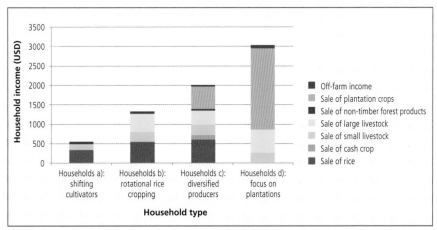

Figure 3: Average annual household incomes by source of income and household type, Viengkhang, Lao PDR. Source: (5)
(N = 504 households in 7 villages)

Diversification – a historical perspective

Muriel Borgeat-Theler

Across decades or even centuries, farmers in many mountain areas have recognized the importance of diversification. Historically, this diversification has taken many avenues, including creating or exploiting sources of income within farming as well as beyond. However, over the course of time, their importance and the need to tap them has changed, as seen in the Valais region of the Swiss Alps.

Tourists with a local guide and the hut warden, around 1900 (© Médiathèque Valais)

For centuries, the Valais, an inner-alpine canton in Switzerland, has been an agrarian region, based on subsistence farming across its different altitudinal belts – from valleys to the alpine zone. Transalpine traffic between Italy and the areas north of the Alps laid an early foundation for an economy based on transit trade. This traffic also offered opportunities to complement incomes from farming, creating new jobs for the supply and maintenance of carts, the storage and guarding of goods, and the provision of food and lodging for travellers (Figure 1).

By the beginning of the fourteenth century, the population had grown to such an extent that the ability to produce food had reached its limits. Widespread poverty was worsened in 1349, when an outbreak of plague killed about one-third of the population. When this decrease in population, and thus the number of mouths to feed, freed up land previously used for growing cereal, the wealthier farmers converted it to hay meadows. This enabled them to increase their cattle herds in order to meet the growing demand from the urban centres of northern Italy.

The rise in cattle farming led to the striking expansion of *bisses* (irrigation channels) in the first half of the fifteenth century, which brought water from the mountains to irrigate the pastures for increasing their productivity. By the beginning of the nineteenth century, the Valais had more than 200 *bisses* with a total length of 2,000 km (1). The descendants of the families that grew rich through cattle farming and trade took over the positions of power. The most famous among them was Kaspar Jodok Stockalper (1609–1691), whose palais was the largest secular building in Switzerland at that time.

At the same time, military service abroad, especially with France and Spain, offered another option for earning an income. In the seventeenth and eighteenth centuries, about 1% of the male population left for military service abroad every year (2).

While first taken up by aristocratic families, a large number of peasants also enrolled. A simple soldier could not make a fortune, and dire poverty was not enough reason to decide to leave his home – it turned out that going off with a bunch of comrades with the chance to indulge in debauchery was a reason just as important as the opportunity to earn some meagre pay. Foreign service was finally banned in 1859.

In the nineteenth century, a steadily increasing population led many families to emigrate. Due to frequent flooding of the Rhone, the region's main river, the government built dykes to contain the river and, as a result, expanded the crop-growing area, and orchards and market gardening spread in the plain. The engineering works themselves created substantial local employment, as did the construction of the railway at about the same time. When tourists discovered the mountains, hotels were built. For the mountain farmers and their families, this had many positive aspects: jobs for local people, the development of guide and driving services, and the opening of various kinds of shops. In the valleys, as industry took off following the arrival of the railway, farmers could combine farming with factory work.

In spite of these developments, society remained primarily agricultural until after the Second World War, when the younger generation turned its back on subsistence agriculture and abandoned the mountain villages. Between 1950 and 1970, the proportion of farmers in the working population fell from 42 to 15%. Although most people in the Valais now work in industry and services, they still have an emotional attachment to the land, and many have a supplementary agricultural activity, such as vine growing or breeding local cattle that they take up to the mountain pastures every summer.

Lessons learned

- In a historical perspective, opportunities for diversification of mountain livelihoods have largely arisen due to demand from outside the mountains.
- Opportunities have changed with the historical context and with local capacities for uptake.
- Local employment opportunities in industry and services help avoid emigration, but not necessarily the abandonment of family farming – as the second half of the twentieth century has shown.

Maintaining the irrigation channel of Savièse, around 1910 (© J. Lüscher, Médiathèque Valais)

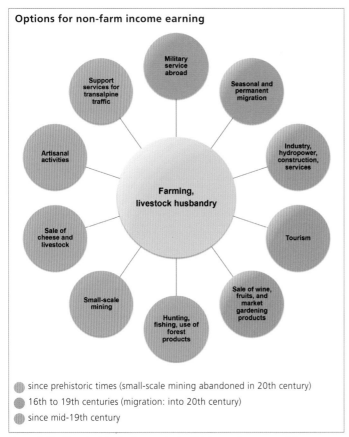

Figure 1: Livelihoods of family farmers in the Valais over the course of time
Source: Muriel Borgeat-Theler 2013

Small forest-based enterprises reconcile conservation and development

The most famous inhabitants of the Bwindi Impenetrable National Park and World Heritage Site in the Kigezi Highlands of Uganda are its 400 mountain gorillas, almost half of the world's endangered mountain gorilla population (1). The farmers who live on the perimeter of this park have benefitted from a change in government policy about improved access to park resources and a development project focused on livelihoods.

Bwindi Village Walk guide with tourists (R. Faidutti)

Adjacent to the Bwindi forests is one of the most densely populated agricultural regions of Uganda with a population of 300,000 and densities up to 400 people per km². The local population, largely made up of small family farmers, is among the poorest in Uganda. Some 90% depend on rainfed mixed farming (2), and land fragmentation and population pressure are high (3). Farmers have traditionally used a wide range of forest products such as weaving and building materials, medicinal plants, bushmeat, honey, minerals and timber to complement subsistence farming. Before the national park status was declared in 1991, the local communities had free access to these resources, which, for some, provided the only means of subsistence. Once declared a national park, the people who lived in the forest were displaced, and local communities were forbidden to remove forest products. Conflicts between local communities and park staff were inevitable. An act of arson in 1992 burned 5% of the park.

In response to these conflicts, Uganda established Multiple Use Zones (MUZs) to allow communities limited access to national park resources. As not all communities were given such zones, FAO and the United Nations Foundation established the Community-Based Commercial Enterprise Development Project for the Conservation of Biodiversity at Bwindi World Heritage Site. Implemented between 2001 and 2004, it helped provide the family farmers adjacent to Bwindi with alternative livelihoods in line with the regulations of the national park.

The project aimed to improve local capacities to develop and manage natural resource-based local enterprises. Using participatory market analysis and development (MA&D), community members were able to select the most promising

Lessons learned

- The Bwindi example shows that mountain forests provide countless products that are used by family farmers for subsistence and commercial purposes. A key for forest conservation are sustainable livelihoods of the population in the areas adjacent to the forests.

- The Bwindi World Heritage Site continues to attract support from local and international NGOs. There is also strong political support for its long-term conservation. Both kinds of support are important for conserving the site while improving the livelihoods of family farmers in the surrounding areas.

Growing mushrooms with local technology (R. Faidutti)

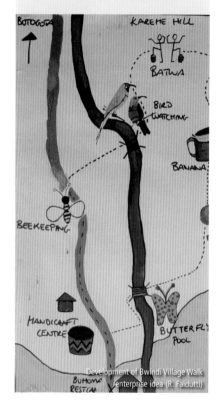

Development of Bwindi Village Walk enterprise idea (R. Faidutti)

products and services, develop business plans and establish viable enterprises. Entrepreneurs examined the entire business environment, including its socio-cultural, environmental, technical, economic and legal aspects. Participating communities were selected by a local government committee that considered the communities' previous engagement in entrepreneurial activities, involvement in forest-based activities, and existing initiatives to form interest groups. Potential entrepreneurs were identified by MA&D facilitators in consultation with the community. The project worked directly with approximately 600 farming households (4).

As a result, 13 enterprise groups were established with about 300 enterprises, including activities in ecotourism, beekeeping, handicrafts, fruit and vegetable cultivation, and support services such as access to credit. The economic contribution of these enterprises to individual households has been notable. For example, traditional handicrafts that now use raw materials such as grasses, grains and dyes grown in home gardens rather than in the park generate an average of USD 17 of additional income per household per month, which is almost 20% of the average monthly income of USD 77 (5). All activities generate employment and additional income, have less impact on land resources than farming and use local knowledge and resources. The majority of the enterprises established are still running successfully (Box). The active involvement of the community from the outset proved key to the success of the project.

The Bohuma Village Walk

This walk, an ecotourism service, was developed by the village of Bohuma with the idea of creating an alternative source of income for farming families and providing an additional offer for tourists visiting the nearby Bwindi National Park to see its mountain gorillas. The walk starts and ends at the entrance of the park and passes through the village, showcasing traditional homesteads, a local women's handicraft centre, a waterfall, tree plantations, a bird-watching spot, and includes a musical performance. The walk has been a great success, with 2,295 visitors in the first 2½ years, and an annual income of USD 6,885 shared between 55 guides and site managers (4,6). In 2006, income increased to USD 13,163 (7). The walk also benefits other community members whose products (e.g. handicrafts) are purchased during the tours. It has also increased the interaction between local population and foreign visitors, and contributes to the conservation of local cultural and natural resources.

(www.bwindiforestnationalpark.com/nature-walk.html)

Social agriculture as part of green care

Christian Hoffmann and Thomas Streifeneder

Green care refers to offering patient-oriented activities that promote physical and mental health and well-being through contact with nature. It recognizes the importance of "social agriculture" activities worldwide. The farms provide various services, such as caretaking, rehabilitation, therapy, education and health care, based on the interplay between people and nature [1]. Offering green care has also proven an innovative agricultural diversification strategy for family farms in the mountain regions of South Tyrol and Trentino in northern Italy.

Schoolchildren learning to deal with animals (Association of South Tyrolean Farm Wives; Andergassen)

In the late 1990s, social agriculture was used in South Tyrol and Trentino to integrate alternative caretaking and home care into daily agricultural activities. The pioneers were farmers who took care of disabled people or people with social or psychological issues at their farms as a matter of course and sometimes even without payment. Today, social agriculture is taking shape as an enterprise that helps create additional income and keeps the farming family employed at home.

In South Tyrol and Trentino, social farming has only been undertaken by a relatively small number of farmers. Though social services provided on farms increased by 83% in the region between 2007 and 2013, the rate was lower than in other European countries, such as Norway and the Netherlands, and compared with other northern Italian regions (2) (Figure 1). Social agriculture is not limited to mountain areas, but the abundance of family farms in mountains, their close relation to nature, focus on animal breeding and traditional farm management make mountain farms advantageous sites for green care activities.

Educational services, such as providing school-on-farm student excursions, kindergarten and childminding, dominate while therapeutic and integrative services are less widespread. The Association of South Tyrolean Farm Wives initiated educational programmes and promoted social agriculture, as a way to combine farm work and social engagement while generating additional income. Students at farm schools pay USD 16 per farm visit, of which the school pays USD 10 and parents pay USD 5, with some 3,250 children visiting in South Tyrol annually. Service-providing farmers must attend special education programmes including a 450-hour course that addresses theory as well as practical work. They must also invest in adapting or building facilities to conform to legal and formal standards. Although public funds are available to cover part of the outlay, the remaining costs

Schoolchildren bake bread (Department of Education on Agriculture, Forest and Domestic Economy)

Lessons learned

- Social agriculture is a promising area of diversification for family farms in mountain areas.
- The required investments and costs of complying with regulations are only affordable if farmers receive direct public compensation or income from institutions financing these farm activities.
- Decision-makers and policy-makers should be made more aware of the need for legal and administrative policies that promote social farm services to complement the services of professional health care providers.
- Farmers who plan to engage in social agriculture must be sure that the farm site, structure and staff are capable of managing the administrative needs and psychological and physical stress.
- Pedagogic, therapeutic or integrative services offered by family farms are promising innovative diversification strategies that can yield inclusive economic growth.

are a challenge for many farmers who wish to engage in social agriculture. In addition to the financial and legal aspects, other fundamental prerequisites to practise green care agriculture have to be fulfilled, including personal interest and skills, education, time and availability and accessibility of the farm facilities.

Although the scope and potential are wide, social agriculture is not an alternative to professional medical-care or caretaking institutions, and research is needed to determine when social agriculture may best be used as complementary treatment. Certified course programmes to meet the legal requirements for these professional services are available in South Tyrol for for childminders, farm school managers, as well as for geriatric care.

> **Social agriculture** is part of green care. Farms following this diversification option fit perfectly in the entrepreneurial concept of multifunctional farming. In addition to their agricultural activities, farmers engaged in social agriculture offer therapeutic, pedagogic or integrative services, which are complementary to professional health care or educational services. Engagement in social agriculture provides additional income to farmers and appreciation alongside agricultural work (3, 4).

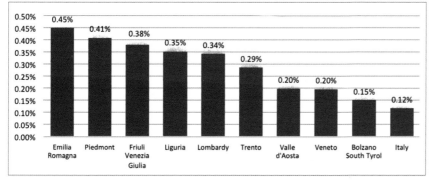

Figure 1: Percentage of "pedagogic farms" in northern Italian regions (2)

Taking care of elderly people from the region (Department of Education on Agriculture, Forest and Domestic Economy)

Rural tourism promotion builds on local values

Katarzyna Sliwa-Martinez

Farmers in the Carpathian Mountains of Poland struggle with economic and structural problems as well as with increasing food demand. At the same time, tourism offers new possibilities for mountain communities rich in culture and nature. Non-governmental organizations have supported mountain farmers to diversify their livelihoods and increase their incomes through sustainable rural tourism.

Shepherd practising traditional transhumance (J. Michalek)

The Carpathian Mountains have been home to rural populations for centuries. On the Polish side of the Carpathians, 70% of the villages still try to make a living with small-scale farming, with an average farm size of 3.5 ha. Challenges in farming are many and include price volatility and demand for agricultural products; migration of young people to the cities; and lack of investment. Tourism offers an opportunity to diversify income, create new jobs for young people and improve the social and economic infrastructure in the villages. Also, there is growing demand for attractive rural and nature-based tourism: The number of agrotourism accommodations has grown by 5–7% per year from 2000 to 2007.

A common feature of farming culture in the Carpathians is the tradition of transhumance – moving cattle, sheep, goats and horses over long distances according to the season. In line with this tradition, the Foundation of Transhumance Pastoralism, together with partners from the Czech Republic, Romania and Slovakia, organizes a migration of 300 sheep through the Carpathians every year. For 100 summer days, experienced shepherds lead multiple flocks of sheep across 1,200 km of rugged terrain, and hold celebratory events with local music, products and handicrafts. These events have become a great tourist attraction. Many farmers have returned to the production of local cheese, which they sell on these occasions. The project also fostered a renewal of pastoral identity and brought new appreciation for the work of the traditional senior shepherds, the *baca*.

Good-quality accommodations in traditional Carpathian style attract tourists – and draw them to stay for a longer period of time. Unfortunately, the standard of accommodations in many rural areas is not very high and there is a lack of tourist

"*We have decided that in our house the most important things are culture, nature and a homey and friendly atmosphere. That is why we offer our guests activities that are based on the traditions and nature of our Carpathian region.*"

Owner of an award-winning rural guest house

People and animals on the way, Zakopane, Poland (J. Michalek)

information and no booking system. Therefore, Pro Carpathia, an association for the development and promotion of the Carpathian region, introduced the Green Tourism Certification scheme in 2010. It certifies four categories of rural services: accommodation, gastronomy, educational services and ecotourism products. In order to qualify for the certification, the accommodation must offer activities such as hiking, Nordic walking, snowshoe trekking, biking, handicrafts and cooking workshops to make and use local products. The certification scheme also looks at the accessibility of information for tourists, standards and quality, effective use of resources, friendliness to children, disabled guests and animals, as well as environmental friendliness.

Sustainable rural tourism continues to grow, engaging new players with new offers: the Polish Society for the Protection of Birds has selected 12 rural townships to prepare an environmentally friendly development strategy, including tourism, with the long-term vision of improving local people's quality of life while protecting Carpathian birds. This initiative is important because long-term planning and community engagement is a prerequisite for sustainable tourism development in rural areas.

Lessons learned

- Good-quality accommodation linked to local tradition and culture is essential for rural mountain tourism development.
- Alongside innovative individual farmers, rural mountain tourism development also needs collective action; examples include the creation of tourist information centres, stepping up local farming and food production to meet tourist demand, local initiatives to build a regionspecific tourist portfolio including innovative offers such as wine tourism or active tourism for seniors and families.

Folk dancers from Poland (M. Borowczyk)

The future of family farming in mountains: policy messages

Schoolchildren in Lao PDR esteem the value of agrobiodiversity in a role play (S. Wymann von Dach)

Credit circle, Myanmar (T. Kohler)

Thomas Kohler, Rosalaura Romeo, Alessia Vita, Maria Wurzinger, Susanne Wymann von Dach

The future of family farming in mountains: policy messages

Family farming, the main type of land use in mountain regions worldwide, is practised in many different ways. Mountain environments require specific management and husbandry choices to cope with altitudinal or seasonal changes in resource availability. Mountain farmers have developed specific techniques, institutions and knowledge which enable them to make a living in mountain environments.

The benefits of family farming in mountains go far beyond the mountain regions and contribute to societies at large. Globally, mountains provide freshwater to half of the world's population, in many ways thanks to the water and soil management practised by mountain farmers. Mountains are also reservoirs of global biodiversity including agro-biodiversity – and mountain farmers have been the custodians of this genetic richness for centuries.

Yet today, family farming in mountain regions is undergoing rapid transformation, due to both internal and external drivers such as population growth, economic globalization and market integration, penetration of urban lifestyles, outmigration of men and youth, and the resulting increased workload for women who remain behind, and increasing claims on land for conservation and large-scale resource extraction, such as mining. These have contributed to higher pressure on local resources, unsustainable practices in land use, disintegration of local customs and traditions, and increased vulnerability to global change.

At the same time, however, these drivers of transformation can also provide opportunities for local development, enhancing the role of family farming and improving the quality of life of mountain farmers. For example, they offer opportunities for increasing farm production sustainably and for diversifying livelihoods by engaging in non-farm activities such as tourism and marketing of local handicrafts. The case studies presented in this publication, collected from mountain regions around the world, show how mountain regions and family farmers benefit from the opportunities these social, economic and environmental transformations present.

Policy messages

- Mountain family farmers and pastoralists have always adapted actively to change, and they continue to do so. However, these efforts need to be supported by enabling policies that will help them adapt to ongoing changes in a sustainable way, in order to achieve sustainable livelihoods and maintain important mountain ecosystem services for themselves and for the many people living downstream.

- National policies that support secure land tenure, access to resources and empower women are a key requirement for promoting sustainable family farming in mountain regions. The same is true for public investment in education, health, transport and research, and for extension services that support farmers in achieving sustainable farming practices through advice in areas such as appropriate use of external inputs including seeds, fertilizers and pesticides.

- Access to credit is crucial for mountain farmers. Their specific conditions call for special granting criteria, including providing access to credit without collateral.

Appropriate technology (T. Kohler)

- Good practices in mountain farming over time and across countries need to be preserved and disseminated. Inappropriate techniques used in mountain environments can quickly lead to erosion, land degradation and even desertification. Innovative techniques and traditional knowledge need to be carefully integrated to increase and restore resilience, with the promotion of a repository of successful practices to be shared globally.

- Production for home consumption is now and will remain the main aim of most family farms in mountains, especially in developing countries. However, better access to markets and credit, and equal rights for women and men can help farmers optimize allocation of farm resources, increase their income and move out of poverty. Creating, labelling and selling quality mountain products derived from organic production is one method farmers can use to improve their livelihoods. Encouraging collaboration and implementing activities such as farmer associations and cooperatives can help lower the barriers to access the markets.

- Policies for mountain family farming need to be embedded in an overall policy of regional mountain development. These policies should promote regional centres and small towns, which provide alternative employment opportunities in the artisanal, industrial and service sectors, stimulate the local economy and reduce outmigration – especially youth outmigration, which is a serious problem in many mountain areas.

- Sustainable mountain family farming produces ecosystem services that are vital for downstream areas and for which farmers should be compensated. Such services include sound watershed management for the provision of freshwater, conservation of biodiversity including vital genetic resources of locally adapted crops and livestock, and attractive cultural landscapes for tourism and recreation.

- Context-specific policies must be shaped for promoting family farming in mountain regions. In the low and middle income countries, promotion should be coupled with poverty alleviation policies. Urbanization, market integration and infrastructure development, which drive development, hold risks as well as opportunities for mountain farmers. The high income countries face the increasing loss of family farms in mountains, which brings with it the real possibility that this type of family farming may disappear completely, taking with it an important element of mountain cultural heritage.

Authors and editors

1 Mountain farming is family farming

Thomas Kohler. University of Bern, Centre for Development and Environment (CDE). Bern, Switzerland. thomas.kohler@cde.unibe.ch

Rosalaura Romeo. Mountain Partnership Secretariat, Food and Agriculture Organization of the United Nations (FAO). Rome, Italy. rosalaura.romeo@fao.org

2 Global change and mountain livelihoods

Hans Hurni. University of Bern, Centre for Development and Environment (CDE). Bern, Switzerland. hans.hurni@cde.unibe.ch

Transformation of mountain livelihoods

Nakileza Bob Roga. Makerere University, Mountain Resource Centre. Kampala, Uganda. nakilezqab@yahoo.com

Mukwaya Peter. Makerere University, Geography, Geo-informatics and Climate Science. Kampala, Uganda. mukwaya@gmail.com

Crisis offers chances for tourism and organic farming

Jelena Krivcevic. Regional Development Agency Bjelasica, Komovi & Prokletije. Berane, Montenegro. jkrivcevic@bjelasica-komovi.co.me

Between melting glaciers, a growing metropolis and the world market

Dirk Hoffmann. Bolivian Mountain Institute (BMI). La Paz, Bolivia. dirk.hoffmann@bolivian-mountains.org

Liz Lavadenz. Bolivian Mountain Institute (BMI). La Paz, Bolivia. liz.lavadenz@bolivian-mountains.org

Rodrigo Tarquino. Centro de Análisis Espacial – Instituto de Ecología UMSA. La Paz, Bolivia. rodrigo.tarquino@gmail.com

Farming on the fringe: adaptation to urbanization

Andreas Haller. University of Innsbruck, Institute of Geography. Innsbruck, Austria. andreas.haller@uibk.ac.at

Oliver Bender. Austrian Academy of Science, Institute for Interdisciplinary Mountain Research. Innsbruck, Austria. oliver.bender@oeaw.ac.at

3 Learning and cooperation

Jill M. Belsky. University of Montana, College of Forestry and Conservation. Missoula, USA. jill.belsky@umontana.edu

Building on traditional cooperation among women

Paolo Ceci. University of Tuscia, Department of Innovation of Biological Systems, Agro-food and Food and Agriculture Organization of the United Nations (FAO). Rome, Italy. paolo.ceci@fao.org

Fatoumata Binta Sombily Diallo. Gamal Abdel Nasser University of Conakry, Centre for Environmental Research (CERE). Conakry, Guinea. binetasombily@yahoo.fr

Petra Wolter. Food and Agriculture Organization of the United Nations (FAO), Forestry Department. Rome, Italy. petra.wolter@fao.org

Lavinia Monforte. Food and Agriculture Organization of the United Nations (FAO), Forestry Department. Rome, Italy. lavinia.monforte@fao.org

A farmers' cooperative and a supermarket team up

Markus Schermer. University of Innsbruck, Mountain Agriculture Research Unit, Department of Sociology. Innsbruck, Austria. markus.schermer@uibk.ac.at

Christoph Furtschegger. University of Innsbruck, Mountain Agriculture Research Unit, Department of Sociology. Innsbruck, Austria. christoph.furtschegger@uibk.ac.at

Radio Mampita – the powerful voice of rural people

Felicitas Bachmann. University of Bern, Centre for Development and Environment (CDE). Bern, Switzerland. felicitas.bachmann@cde.unibe.ch

A school for promoters of agro-ecology

Francisco Medina. Ministry of Environment Peru, UNDP and GEF, Sustainable Land Management Project Apurímac. San Isidro, Peru. fmedina@minam.gob.pe

Jenny Chimayco Ortega. Ministry of Environment Peru, UNDP and GEF, Sustainable Land Management Project Apurímac. San Isidro, Peru. jchimayco@minam.gob.pe

Field schools for agro-pastoralists

Caterina Batello. Food and Agriculture Organization of the United Nations (FAO), Plant Production and Protection Division (AGP). Rome, Italy. caterina.batello@fao.org

James Okoth. Food and Agriculture Organization of the United Nations (FAO), Plant Production and Protection Division (AGP). Rome, Italy. james.okoth@fao.org

Monica Petri. Food and Agriculture Organization of the United Nations (FAO), Plant Production and Protection Division (AGP). Rome, Italy. monica.petri@fao.org

Manuela Allara. Food and Agriculture Organization of the United Nations (FAO), Plant Production and Protection Division (AGP). Rome, Italy. manuela.allara@fao.org

Lobbying for mountain regions and farming

Jörg Beck. Schweizerische Arbeitsgemeinschaft für das Berggebiet (SAB). Bern, Switzerland. joerg.beck@sab.ch

4 Sustainable intensification and organic farming

Maria Wurzinger. BOKU – University of Natural Resources and Life Sciences, Centre for Development Research. Vienna, Austria. maria.wurzinger@boku.ac.at

Urs Niggli. Research Institute of Organic Agriculture. Frick, Switzerland. urs.niggli@fibl.org

Towards a fully organic state

Niraj Nirola. Indian Institute of Technology, Department of Humanities & Social Sciences. Bombay, India. niraj_nirola@hotmail.com

Trilochan Pandey. World Learning / School for International Training (SIT), Sustainable Development and Social Change Program. Jaipur (Rajasthan), India. trilochan.pandey@gmail.com

Kitchen gardens for improved well-being

Elbegzaya Batjargal. University of Central Asia, Mountain Partnership Secretariat – Decentralized Hub for Central Asia. Bishkek, Kyrgyz Republic. elbegzaya.batjargal@ucentralasia.org

Tamana Zamir. Aga Khan Development Network, Kyrgyzstan Mountain Societies Development Support Programme (MSDSP KG). Bishkek, Kyrgyz Republic. tamana.zamir@akdn.org

Organic farming improves income and diet

Gerhard Buttner. Christian Aid. Lima, Peru. gbuttner@christian-aid.org

Cecilia Gianella. Christian Aid. Lima, Peru. cgianella@christian-aid.org

Sustainable mountain pastoralism: challenges and opportunities

Henri Rueff. University of Oxford, School of Geography and the Environment, and University of Bern, Centre for Development and Environment (CDE). Oxford, United Kingdom. henri.rueff@geog.ox.ac.uk

Inam-ur-Rahim. Foundation for Research and Socio-Ecological Harmony. Islamabad, Pakistan. irahim33@yahoo.com

Improvement of aquaculture practices in mountain farming

Nguyễn Hữu Nghĩa. Research Institute for Aquaculture No 1. Dinh Bang, Tu Son, Viet Nam. nghia@ria1.org

Hoang Thu Thuy. Research Institute for Aquaculture No 1. Dinh Bang, Tu Son, Viet Nam. hoangthuy@ria1.org

Dang Thi Oanh. Research Institute for Aquaculture No 1. Dinh Bang, Tu Son, Viet Nam. dtoanh@ria1.org

Tran Minh Hau. Research Institute for Aquaculture No 1. Dinh Bang, Tu Son, Viet Nam. tmhau@ria1.org

Organic farming as a climate change adaptation measure

Carla Marchant Santiago. University of Innsbruck, Geography Department, Interdisciplinary Mountain Research (IGF). Innsbruck, Austria. carla.marchant-santiago@uibk.ac.at

Axel Borsdorf. University of Innsbruck, Geography Department, Interdisciplinary Mountain Research (IGF). Innsbruck, Austria. axel.borsdorf@oeaw.ac.at

5 Mountain products and market development

Susanne Wymann von Dach. University of Bern, Centre for Development and Environment (CDE). Bern, Switzerland. susanne.wymann@cde.unibe.ch

Certification frameworks for mountain products

Markus Schermer. University of Innsbruck, Mountain Agriculture Research Unit, Department of Sociology. Innsbruck, Austria. markus.schermer@uibk.ac.at

Agribusiness development through cooperation

Dyutiman Choudhary. International Centre for Integrated Mountain Development (ICIMOD). Kathmandu, Nepal. dchoudhary@icimod.org

Mahendra Singh Kunwar. Himalayan Action Research Centre (HARC). Dehra Dun, India. info@harcindia.org

Adding value to traditional mountain crops

Alessandra Giuliani. Bern University of Applied Sciences, School for Agricultural, Forest and Food Sciences (HAFL). Zollikofen, Switzerland. alessandra.giuliani@bfh.ch

Stefano Padulosi. Bioversity International. Rome, Italy. s.padulosi@cgiar.org

Spinning a fine yarn

Barbara Rischkowsky. International Center for Agricultural Research in the Dry Areas (ICARDA), Ethiopia Office. Addis Ababa, Ethiopia. b.rischkowsky@cgiar.org

Liba Brent. Consultant. Madison, USA. libabrent@gmail.com

Community-based beekeeping for better livelihoods

Uma Partap. International Centre for Integrated Mountain Development (ICIMOD). Kathmandu, Nepal. upartap@icimod.org

Min B. Gurung. International Centre for Integrated Mountain Development (ICIMOD). Kathmandu, Nepal. mgurung@icimod.org

6 Diversification of mountain livelihoods

Thomas Kohler. University of Bern, Centre for Development and Environment (CDE). Bern, Switzerland. thomas.kohler@cde.unibe.ch

Kata Wagner. Food and Agriculture Organization of the United Nations (FAO), Forestry Department. Rome, Italy. kata.wagner@fao.org

Diversification – a historical perspective

Muriel Borgeat-Theler. Foundation for Sustainable Development of Mountain Regions. Sion, Switzerland. muriel.borgeat@fddm.vs.ch

Small forest-based enterprises reconcile conservation and development

Kata Wagner. Food and Agriculture Organization of the United Nations (FAO), Forestry Department. Rome, Italy. kata.wagner@fao.org

Social agriculture as part of green care

Christian Hoffmann. European Academy of Bozen/Bolzano, Institute for Regional Development and Location Management. Bozen/Bolzano, Italy. christian.hoffmann@eurac.edu

Thomas Streifeneder. European Academy of Bozen/Bolzano, Institute for Regional Development and Location Management. Bozen/Bolzano, Italy. Thomas.Streifeneder@eurac.edu

Rural tourism promotion builds on local values

Katarzyna Sliwa-Martinez. Jagiellonian University. Krakow, Poland. katarzyna.d.sliwa@gmail.com

7 The future of family farming in mountains: policy messages

Thomas Kohler. University of Bern, Centre for Development and Environment (CDE). Bern, Switzerland. thomas.kohler@cde.unibe.ch

Rosalaura Romeo. Mountain Partnership Secretariat, Food and Agriculture Organization of the United Nations (FAO). Rome, Italy. rosalaura.romeo@fao.org

Alessia Vita. Mountain Partnership Secretariat, Food and Agriculture Organization of the United Nations (FAO). Rome, Italy. alessia.vita@fao.org

Maria Wurzinger. BOKU – University of Natural Resources and Life Sciences, Centre for Development Research. Vienna, Austria. maria.wurzinger@boku.ac.at

Susanne Wymann von Dach. University of Bern, Centre for Development and Environment (CDE). Bern, Switzerland. susanne.wymann@cde.unibe.ch

Editors

Susanne Wymann von Dach. University of Bern, Centre for Development and Environment (CDE). Bern, Switzerland. susanne.wymann@cde.unibe.ch

Rosalaura Romeo. Mountain Partnership Secretariat, Food and Agriculture Organization of the United Nations (FAO). Rome, Italy. rosalaura.romeo@fao.org

Alessia Vita. Mountain Partnership Secretariat, Food and Agriculture Organization of the United Nations (FAO). Rome, Italy. alessia.vita@fao.org

Maria Wurzinger. BOKU-University of Natural Resources and Life Sciences, Centre for Development Research. Vienna, Austria. maria.wurzinger@boku.ac.at

Thomas Kohler. University of Bern, Centre for Development and Environment (CDE). Bern, Switzerland. thomas.kohler@cde.unibe.ch

References and further reading

Numbered references are cited in the chapters. All websites were accessed on 17 November 2013.

1 Mountain farming is family farming

(1) Crowley E. 2013. With appropriate support family farming can contribute to the future of sustainable rural development. Rural 21 – The International Journal for Rural Development. www.rural21.com/english/points-of-view/detail/article/family-farming-the-backbone-of-sustainable-rural-development-0000794

(2) FAO [Food and Agriculture Organization]. 2011. Why invest in sustainable mountain development? Rome, Italy: FAO.

(3) FAO [Food and Agriculture Organization]. 2013. IYFF [International Year of Familiy Farming] concept note [in preparation]. Rome, Italy: FAO.

(4) IAASTD [International Assessment of Agricultural Knowledge, Science, and Technology for Development]. 2009. Agriculture at a Crossroad: IAASTD Global Report. Washington DC, USA: Island Press.

2 Global change and mountain livelihoods

(1) Wiesmann U, Hurni H (eds), with an international group of co-editors. 2011. Research for Sustainable Development: Foundations, Experiences, and Perspectives. Perspectives of the Swiss National Centre of Competence in Research (NCCR) North-South, University of Bern, Vol. 6. Bern, Switzerland: Geographica Bernensia.

IAASTD [International Assessment of Agricultural Knowledge, Science, and Technology for Development]. 2009. Agriculture at a Crossroad: IAASTD Global Report. Washington DC, USA: Island Press.

Transformation of mountain livelihoods

(1) Emwanu T, Okiira Okwi P, Hoogeveen JG, Kristjanson P. 2004. Where are the poor? Mapping patterns of well-being in Uganda 1992 and 1999. Entebbe, Uganda: Uganda Bureau of Statistics and ILRI [International Livestock Research Institute].

(2) Mukwaya P, Bamutaze Y, Mugarura S, Benson T. 2012. Rural–urban transformation in Uganda. Working Paper. Kampala, Uganda: Uganda Strategy Support Program and IFPRI [International Food Policy Research Institute].

(3) Templeton SR, Scherr SJ. 1997. Population pressure and the microeconomy of land management in hills and mountains of developing world. Discussion paper 26. EPTD [Environment and Production Technology Division]. Washington DC, USA: IFPRI [International Food Policy Research Institute].

Bernard C, Nakileza B, Mtasiwa B, Mbataru P, Nantumbwe C, Sarr A, Gathoni E, Tilumanywa VT. 2010. Rural–urban linkages and livelihoods: Small projects and farming systems in Mt Elgon [unpublished report to the Collaborative Research on East Africa Territorial Integration within Globalization Program].

Crisis offers chances for tourism and organic farming

(1) Jovanović M, Despotović A. 2012. The analysis of socio-economic conditions for organic production in Montenegro. Economics of Agriculture 2:207–215.

(2) Montenegro Statistical Office. 2012. Poverty analysis in Montenegro in 2011. Release No 329. Podgorica, Montenegro.

(3) Montenegro Ministry of Tourism and Environment. 2008. Montenegro tourism development strategy to 2020. Podgorica, Montenegro.

Between melting glaciers, a growing metropolis and the world market

(1) Sergeotecmin. 2012. State Mining Service. La Paz, Bolivia.

BIO-THAW [Modeling BIOdiversity and land use interactions under changing glacial water availability in Tropical High Andean Wetlands]. 2013. BIOdiversity and people facing climate change in Tropical High Andean Wetlands. www.biothaw.ird.fr

Bury J, Mark BG, Carey M, Young KR, McKenzie JM, Baraer M, French A, Polk MH. 2013. New geographies of water and climate change in Peru: Coupled natural and social transformations in the Santa River Watershed. Annals of the Association of American Geographers 103(2):363–374.

Hoffmann D, Requena C. 2012. Bolivia en un mundo 4 grados más caliente. Escenarios sociopolíticos ante el cambio climático para los años 2030 y 2060 en el altiplano norte. La Paz, Bolivia: Instituto Boliviano de la Montaña and Fundación PIEB.

Sietz D, Mamani Choque SE, Lüdeke MKB. 2012. Typical patterns of smallholder vulnerability to weather extremes with regard to food security in the Peruvian Altiplano. Regional Environmental Change 12(3):489–505.

Soruco Sologuren Á. 2012. Medio siglo de fluctuaciones glaciares en la Cordillera Real y sus efectos hidrológicos en la ciudad de La Paz. La Paz, Bolivia: IRD [Institut de recherche pour le développement].

Farming on the fringe: adaptation to urbanization

(1) Pulgar Vidal J. 1996. Geografía del Perú. Las ocho regiones naturales, la regionalización transversal, la sabiduría ecológica tradicional. Lima, Peru: PEISA.

(2) INEI [Instituo Nacional de Estadística e Informática]. 2012. Perú: Estimaciones y proyecciones de población total por sexo de las principales ciudades, 2000–2015. Boletín Especial No 23. Lima, Peru: INEI.

(3) Haller A, Borsdorf A. 2013. Huancayo Metropolitano. Cities 31:553–562.

(4) Haller A. 2012. Vivid valleys, pallid peaks? Hypsometric variations and rural–urban land change in the Central Peruvian Andes. Applied Geography 35(1–2):439–447.

(5) INEI [Instituo Nacional de Estadística e Informática]. 2013. IV Censo nacional agropecuario 2012. Lima, Peru: INEI.

(6) Ministerio del Ambiente del Perú. 2011. Decreto supremo que establece el Área de Conservación Regional Huaytapallana. El Peruano (21 July):446882–446885.

(7) Sauñi J. 2013. Alpacas al nevado. Reemplazarán a vacas y corderos. Correo Huancayo (19 June):9.

Borsdorf A, Stadel C. 2013. Die Anden. Ein geographisches Porträt. Berlin, Germany: Springer.

Gade DW. 1999. Nature and Culture in the Andes. Madison, USA: University of Wisconsin Press.

Murra JV. 2009. El mundo andino, población, medio ambiente y economía. Lima, Peru: Instituto de Estudios Peruanos.

Perlik M, Kohler T. 2012. Green economy and urbanization in mountains. In: Kohler T, Pratt J, Debarbieux B, Balsiger J, Rudaz G, Maselli D (eds). Sustainable Mountain Development: Green Economy and Institutions. From Rio 1992 to Rio 2012 and beyond. Bern, Switzerland.

Satterthwaite D, McGranahan G, Tacoli C. 2010. Urbanization and its implications for food and farming. Philosophical Transactions of the Royal Society B 365(1554):2809–2820.

3 Learning and cooperation

(1) Ostrom E. 1990. Governing the Commons: The Evolution of Institutions for Collective Action. New York, USA: Cambridge University Press.

(2) Kauffman J. 2013. Conservationists go big in Montana. Land & People. Trust for Public Land. www.tpl.org/magazine/conservationists-go-big-montana-%E2%80%93landpeople

Reed MS, Evely AC, Cundill G, Fazey I, Glass J, Laing A, Newig J, Parrish B, Prell C, Raymond C, Stringer LC. 2010. What is social learning? Ecology and Society 15(4):r1 [online].

Simão Seixas C, Berkes F. 2010. Community-based enterprises: The significance of partnerships and institutional linkages. International Journal of the Commons 4(1):183–212.

Wondolleck JM, Yaffee SL. 2000. Making Collaboration Work: Lessons from Innovation in Natural Resource Management. Washington DC, USA: Island Press.

Building on traditional cooperation among women

(1) FAO [Food and Agriculture Organization]. 2008. Fouta Djallon Highlands Integrated Natural Resources Management Project (FDH-INRM) [internal project document]. Rome, Italy: FAO.

(2) IFAD [International Fund for Agricultural Development]. Rural poverty in Guinea. Rural Poverty Portal. www.ruralpovertyportal.org/country/home/tags/guinea

(3) FAO [Food and Agriculture Organization]. 2013. Draft addendum for tranche II to the project document [unpublished]. Rome, Italy: FAO.

(4) Detraux M. 1991. Approche intégrée des systèmes de production et de leur dynamisme, un outil pour une politique adaptée aux besoins des régions: application au Fouta Djallon. Gembloux, Belgium: Faculté universitaire des sciences agronomiques de Gembloux.

(5) Ceci P. 2013. Interweaving forests into society: Towards long-term impacts and sustainability of forestry projects in Guinea [PhD thesis in preparation]. Viterbo, Italy: University of Tuscia.

A farmers' cooperative and a supermarket team up

Bio vom Berg. www.biovomberg.at

Radio Mampita – the powerful voice of rural people

(1) Bachmann F. et al. [in preparation]. The role of farmer-owned and private radio in rural development. Case studies from Madagascar and Kenya [internal project report]. Bern, Switzerland: Centre for Development and Environment (CDE), University of Bern.

Fraser C, Restrepo Estrada S. 2001. Community Radio Handbook. Paris, France: UNESCO [United Nations Educational, Scientific and Cultural Organization].

Al-hassan S, Andani A, Abdul-Malik A. 2011. The role of community radio in livelihood improvement: The case of Simli Radio. Field Actions Science Reports, Vol. 5 [online].

CIMA [Center for International Media Assistance]. 2007. Community radio: Its impact and challenges to its development. Working Group Report, 9 October 2007. Washington DC, USA.

AMARC [World Association of Community Radio Broadcasters]. www2.amarc.org

A school for promoters of agro-ecology

Ministerio del Ambiente. Proyecto Manejo Sostenible de la Tierra Apurímac. www.minam.gob.pe/mst

Ministerio del Ambiente. 2013. Ecohéroes. La ruta verde de los peruanos del mañana. Proyecto Manejo Sostenible de la Tierra en Apurímac. Lima, Peru: Ministerio del Ambiente.

Proyecto Manejo Sostenible de la Tierra en Apurímac. 2013. Estudio de línea de base biofísica y socioeconómica del ámbito del proyecto [unpublished]. Peru.

Proyecto Manejo Sostenible de la Tierra en Apurímac. 2013. Diseño e implementación de planes de manejo de recursos naturales (pastizales, agroforestería, bosques) que integran beneficios económicos y conservación de servicios ecosistémicos [unpublished study]. Peru.

Proyecto Manejo Sostenible de la Tierra en Apurímac. 2013. Caracterización de los sistemas de manejo comunal de germoplasma en 23 comunidades campesinas [unpublished study]. Peru.

Proyecto Manejo Sostenible de la Tierra en Apurímac. 2013. Estudio de diagnóstico frutícola en las subcuencas de Vilcabamba media y alta y Santo Tomás media [unpublished study]. Peru.

Field schools for agro-pastoralists

(1) Neely C, Bunning S, Wilkes A (eds). 2009. Review of evidence on drylands pastoral systems and climate change – Implications and opportunities for mitigation and adaptation. Land and Water Discussion Paper 8. Rome, Italy: FAO [Food and Agriculture Organization].

(2) Behnke R, Kerven C. 2013. Climate resilience, productivity and equity in drylands. Climate Change Working Paper No 4. London, United Kingdom: IIED [International Institute for Environment and Development].

(3) Shanahan M. 2013. Media perceptions and portrayals of pastoralists in Kenya, India and China. Gatekeeper 154 (April 2013). London, United Kingdom: IIED [International Institute for Environment and Development].

Lipper L, Cavatassi R, Winters PC. 2005. Seed systems, household welfare and crop genetic diversity: An economic methodology applied in Ethiopia. ESA Technical Paper. Rome, Italy: FAO [Food and Agriculture Organization].

Sen A. 2013. Why is there so much hunger in the world? McDougall Memorial Lecture at Food Security Conference on 15 June 2013. Rome, Italy: FAO [Food and Agriculture Organization].

Lobbying for mountain regions and farming

(1) Schweizerischer Bundesrat. 2012. Botschaft zur Weiterentwicklung der Agrarpolitik in den Jahren 2014–2017 (Agrarpolitik 2014–2017). 12.021. Bern, Switzerland. www.admin.ch/opc/de/federal-gazette/2012/2075.pdf

(2) Schweizerische Arbeitsgemeinschaft für das Berggebiet. www.sab.ch

(3) Egger T, Favre G, Passagla M. 2008. Der Agrotourismus in der Schweiz – Analyse der aktuellen Situation und Empfehlungen für die Zukunft. SAB [Schweizerische Arbeitsgemeinschaft für das Berggebiet] No 194. Bern, Switzerland: SAB.

(4) Die Bundesbehörden der Schweizerischen Eidgenossenschaft. 2013. Schweizer Verordnung vom 25. Mai 2011 über die Verwendung der Bezeichnungen «Berg» und «Alp» für landwirtschaftliche Erzeugnisse und daraus hergestellte Lebensmittel (Berg- und Alpverordnung). SR 910.19. Bern, Switzerland.

(5) BAKOM [Bundesamt für Kommunikation] et al. (ed.). 2012. Wege zur Datenautobahn. Hochwertiges Breitband – ein Leitfaden für Gemeinden, Regionen und Kantone. Biel, Switzerland: BAKOM.

(6) BLW [Bundesamt für Landwirtschaft]. 2012. Agrarbericht 2012. Bern, Switzerland: BLW.

4 Sustainable intensification and organic farming

(1) The Royal Society. 2009. Reaping the benefits: Science and the sustainable intensification of global agriculture. London, United Kingdom.

(2) Garnett T, Godfray C. 2012. Sustainable intensification in agriculture: Navigating a course through competing food system priorities. Food Climate Research Network and the Oxford Martin Programme on the Future of Food. Oxford, United Kingdom.

(3) Niggli U, Slabe A, Schmid O, Halberg N, Schlüter M. 2008. Vision for an organic food and farming research agenda 2025: Organic knowledge for the future. Technology Platform Organics. Brussels, Belgium: IFOAM EU Group [International Federation of Organic Agriculture Movements Regional Group European Union] and Bonn, Germany: ISOFAR [International Society of Organic Agriculture Research].

(4) Willer H, Lernoud J, Kilcher L (eds). 2013. The world of organic agriculture: Statistics and emerging trends 2013. Frick, Switzerland: FiBL [Research Institute of Organic Agriculture] and Bonn, Germany: IFOAM [International Federation of Organic Agriculture Movements].

(5) Hine R, Pretty J, Twarog S. 2008. Organic agriculture and food security in Africa. (UNCTAD/DITC/TED/2007/15.) Geneva, Switzerland and New York, USA: UNEP–UNCTAD CBTF [United Nations Environment Programme – United Nations Conference on Trade and Development, Capacity Building Task Force].

Towards a fully organic state

Planning Commission. 2008. Sikkim development report. New Delhi, India: Academic Foundation.

Sharma G, Liang L, Tanaka K, Subba J, Sharma E. 2009. Sikkim Himalayan agriculture: Improving and scaling up of the traditionally managed agricultural systems of global significance. Resources Science 31(9):21–30.

Ramesh P, Panwar NR, Singh AB, Ramana S, Yadav S, Shrivastava R, Subba Rao A. 2010. Status of organic farming in India. Current Science 98(9):1190–1194.

Kitchen gardens for improved well-being

(1) AKF [Aga Khan Foundation]. 2005. Analysis of health and nutrition survey in Chonq-Alai and Alai districts of Osh Oblast 2004 [unpublished study]. Geneva, Switzerland.

(2) MSDSP KG [Mountain Societies Development Support Programme Kyrgyz Republic]. 2011. Kitchen garden project monitoring study [unpublished study]. Bishkek, Kyrgyz Republic.

(3) MSDSP KG [Mountain Societies Development Support Programme Kyrgyz Republic]. 2012. Analysis of health knowledge and practice survey on Chong-Alai and Alai rayons [unpublished study]. Osh, Kyrgyz Republic.

Ministry of Health of Kyrgyz Republic. 2011. Den Sooluk National Health Reform Programme in Kyrgyz Republic 2012–2016. Bishkek, Kyrgyz Republic.

UNICEF [United Nations Children's Fund]. 2011. Survey on nutritional status of children and health awareness in Alai and Chong-Alai rayons of Osh region. Bishkek, Kyrgyz Republic.

Organic farming improves income and diet

Christian Aid in Peru. www.christianaid.org.uk/whatwedo/the-americas/peru.aspx and www.christianaid.org.uk/whatwedo/partnerfocus/cedap-peru.aspx

Christian Aid in Latin America. www.christianaid.org.uk/whatwedo/in-focus/gender/latin_america_caribbean.aspx

Centro de Desarrollo Agropecuario. www.cedap.org.pe

Sustainable mountain pastoralism: challenges and opportunities

(1) Sharma E, Zhaoli Y, Sharma B. 2007. ICIMOD's regional rangeland program for the Hindu Kush–Himalayas. Mountain Research and Development 27:174–177.

(2) Westreicher CA, Mérega JL, Gabriel Palmili G. 2006. Review of the literature on pastoral economics and marketing: South America. Nairobi, Kenya: WISP [World Initiative for Sustainable Pastoralism].

(3) Montero R, Mathieu J, Singh C. 2009. Mountain pastoralism 1500–2000: An introduction. Nomadic Peoples 13:1–16.

(4) Kreutzmann H. 2012. Pastoral practices in transition: Animal husbandry practices in high Asian contexts. In: Kreutzmann H (ed.). Pastoral Practices in High Asia: Agency of Development Effected by Modernization, Resettlement and Transformation. Dordrecht, Netherlands: Springer, pp. 1–29.

(5) Cabral L. 2009. Sector-based approaches in agriculture: Past experience, current setting and future options. London, United Kingdom: Overseas Development Institute.

(6) Rahim I, Maselli D, Rueff H, Wiesmann U. 2011. Indigenous fodder trees can increase grazing accessibility for landless and mobile pastoralists in Northern Pakistan. Pastoralism: Research, Policy and Practice 1(2).

(7) Ojeda G, Rueff H, Rahim I, Maselli D. 2012. Sustaining mobile pastoralists in the mountains of northern Pakistan. Evidence for Policy Series. Regional edition Central Asia, No 3. Bishkek, Kyrgyzstan: NCCR North-South.

(8) Shah I, Rahim I, Rueff H, Maselli D. 2012. Landless mobile pastoralists: Securing their role as custodian of northern Pakistan's mountains. Workshop proceedings. Bern, Switzerland: Centre for Development and Environment (CDE), University of Bern.

(9) WISP [World Initiative for Sustainable Pastoralism]. 2007. Total economic valuation of Kyrgyzstan pastoralism. Nairobi, Kenya: WISP.

(10) Rahim I, Saleem M, Rueff H, Maselli D. 2013. Conserving indigenous livestock breeds to benefit mountain smallholders. Evidence for Policy Series. Regional edition Central Asia, No 6. Ed. by Mira Arynova. Bishkek, Kyrgyzstan: NCCR North-South.

(11) Saleem M, Rahim I, Jalali S, Rueff H, Khan M, Maselli D, Wiesmann U, Muhammad S. 2013. Morphological characterization of Achai cattle in sedentary and transhumant systems in Pakistan. Animal Genetic Resources 52:83–90.

(12) Jan P, Lips M, Dumondel M. 2012. Total factor productivity change of Swiss dairy farms in the mountain region in the period 1999 to 2008. Review of Agricultural and Environmental Studies 93:273–298.

Improvement of aquaculture practices in mountain farming

Research Institute for Aquaculture No 1. Viet Nam. www.ria1.org/ria1/Default.aspx?tab=61&LangID=2

Organic farming as a climate change adaptation measure

(1) Borsdorf A, Mergili M (eds). 2011. Kolumbien im Wandel. Erkenntnisse und Eindrücke einer dreiwöchigen Studienexkursion durch Zentral- und Südkolumbien. inngeo – Innsbrucker Materialien zur Geographie No 14. Innsbruck, Austria: Innsbrucker Studienkreis für Geographie.

(2) Borsdorf A, Borsdorf F, Ortega LA. 2011. Towards climate change adaptation, sustainable development and conflict resolution – the Cinturón Andino Biosphere Reserve in Southern Colombia. eco.mont – Journal on Protected Mountain Areas Research and Management 3(2):43–48.

(3) Borsdorf A, Marchant C, Mergili M (eds). 2013. Agricultura ecológica y estrategias de adaptación al cambio climático en la Cuenca del Río Piedras. Popayán, Colombia: University of Innsbruck.

Austrian MAB Committee (ed.). 2011. Biosphere Reserves in the Mountains of the World. Excellence in the Clouds? Vienna, Austria: Austrian Academy of Sciences Press.

Batisse M. 1997. Biosphere Reserves: A challenge for biodiversity conservation and regional development. Environment: Science and Policy for Sustainable Development 39(5):6–33.

5 Mountain products and market development

(1) Hoermann B, Choudhary D, Choudhury D, Kollmair M. 2010. Integrated value chain development as a tool for poverty alleviation: An analytical and strategic framework. Kathmandu, Nepal: ICIMOD [International Centre for Integrated Mountain Development].

(2) Williams S, Kepe T. 2008. Discordant harvest: Debating the harvesting and commercialization of Wild Buchu (Agathosma betulina) in Elandskloof, South Africa. Mountain Research and Development 28(1):58–64.

(3) Giuliani A, Hintermann F, Rojas W, Padulosi S (eds). 2012. Biodiversity of Andean grains: Balancing market potential and sustainable livelihoods. Rome, Italy: Bioversity International.

(4) Pasca A, Guitton M, Rouby A. Guidelines for the development, promotion and communication of mountain foods. Brussels, Belgium: Euromontana.

(5) CIP [International Potato Centre]. Potato, native species. http://cipotato.org/potato/native-varieties

(6) ICIMOD [International Centre for Integrated Mountain Development]. HKH conservation portal: Species data set. Kathmandu, Nepal. http://hkhconservationportal.icimod.org/MetadataSpecies.aspx

(7) TABI [The Agrobiodiversity Initiative], Ministry of Agriculture and Forestry. 2013. NTFP database [unpublished database]. Vientiane, Lao PDR. www.tabi.la

(8) FAO [Food and Agriculture Organization]. Domestic animal diversity information system. http://dad.fao.org/

(9) Swiss Schabziger (sapsago). Brand product since 1463. www.schabziger.ch/en/brand-product-since-1463

(10) Lüscher E, Frei B. 2013. 550 Jahre Schabziger – Geschichte und Rezepte. Lenzburg, Switzerland: Fona Verlag.

Certification frameworks for mountain products

(1) Euromontana. 2007–2009. EuroMarc – Mountain agrofood products in Europe, their consumers, retailers and local initiatives. www.euromontana.org/en/projets/euromarc.html

(2) Euromontana. European Charter of mountain food products. www.euromontana.org/en/themes-detravail/european-charter-of-mountain-food-products.html

(3) Santini F, Guri F, Gomez y Paloma S. 2013. Labelling of agricultural and food products of mountain farming. JRC [Joint Research Centre] scientific and policy report. Seville, Spain: European Commission, JRC.

(4) The Council of the European Union. 1999. Council Regulation (EC) No 1257/1999. http://eur-lex.europa.eu/LexUriServ/LexUriServ.do?uri=OJ:L:1999:160:0080:0102:EN:PDF

(5) European Union. 2012. Regulation (EU) No 1151/2012 of the European Parliament and of the Council on quality schemes for agricultural products and foodstuff. Official Journal of the European Union 14.12.2012, L343/1-29. http://eur-lex.europa.eu/LexUriServ/LexUriServ.do?uri=OJ:L:2012:343:0001:0029:en:PDF

(6) Pasca A, Guitton M, Rouby A. 2010. Designation and promotion of mountain quality food products in Europe: Policy recommendations. Brussels, Belgium: EuroMarc.

Agrobusiness development through cooperation

(1) Choudhary D, Ghosh I, Chauhan S, Bhati S, Juyal M. 2013. Case studies on value chain approach for mountain development in Uttarakhand, India. Working Paper 2013/6. Kathmandu, Nepal: ICIMOD [International Centre for Integrated Mountain Development].

(2) Choudhary D, Pandit BH, Kinhal G, Kollmair M. 2011. Pro poor value chain development for high value products in mountain regions: Indian bay leaf. Kathmandu, Nepal: ICIMOD [International Centre for Integrated Mountain Development].

Bolwig S, Ponte S, du Toit A, Riisgaard L, Halberg N. 2010. Integrating poverty and environmental concerns into value-chain analysis: A conceptual framework. Development Policy Review 28(2):173–194.

Hoermann B, Choudhary D, Choudhury D, Kollmair M. 2010. Integrated value chain development as a tool for poverty alleviation: An analytical and strategic framework. Kathmandu, Nepal: ICIMOD [International Centre for Integrated Mountain Development].

Mitchell J, Shepherd A, Keane J. 2011. An introduction. In: Mitchell J, Coles C (eds). Markets and Rural Poverty: Upgrading in Value Chains. London, United Kingdom and Washington DC, USA: Earthscan, IDRC [International Development Research Centre], pp. 217–234.

Adding value to traditional mountain crops

(1) Canahua A, Valdivia R, Mújica A, Allasi M. 2003. Beneficios nutritivos y formas de consumo de la quinua (Chenopodium quinoa Willd) y de la kañihua (Chenopodium pallidicaule Aellen). Puno, Peru: IPGRI-IFAD, CARE-PERU, UNA/II & CIRNMA.

(2) Giuliani A, Hintermann F, Rojas W, Padulosi S (eds). 2012. Biodiversity of Andean grains: Balancing market potential and sustainable livelihoods. Rome, Italy: Bioversity International.

(3) Rojas W, Soto JL, Pinto M, Jäger M, Padulosi S (eds). 2010. Granos Andinos. Avances, logros y experiencias desarrolladas en quinua, cañahua y amaranto en Bolivia. Rome, Italy: Bioversity International.

(4) Padulosi S, Bala Ravi S, Rojas W, Valdivia R, Jager M, Polar V, Gotor E, Bhag Mal. 2013. Experiences and lessons learned in the framework of a global UN effort in support of neglected and underutilized species. ISHS [International Society for Horticultural Science] Acta Horticulturae 979:517–531.

Padulosi S, Bergamini N, Lawrence T (eds). 2012. On farm conservation of neglected and underutilized species: Status, trends and novel approaches to cope with climate change. Proceedings of an International Conference, Frankfurt, Germany, 14–16 June 2011. Rome, Italy: Bioversity International.

Gruèr G, Giuliani A, Smale M. 2009. Marketing underutilized species for the benefit of the poor: A conceptual framework. In: Agrobiodiversity, Conservation and Economic Development. London, United Kingdom and New York, USA: Routledge, pp. 62–81.

Spinning a fine yarn

Ansari-Renani HR, Rischkowsky B, Mueller JP, Seyed Momen SM, Moradi S. 2013. Nomadic pastoralism in southern Iran. Pastoralism: Research, Policy and Practice 3(1):11.

Kosimov FF, Kosimov MA, Mueller JP, Rischkowsky B. 2013. Evaluation of mohair quality in Angora goats from the Northern dry lands of Tajikistan. Small Ruminant Research 113(1):73–79.

Ansari-Renani HR, Mueller JP, Rischkowsky B, Seyed Momen SM, Alipour O, Ehsani M, Moradi S. 2012. Cashmere quality of Raeini goats kept by nomads in Iran. Small Ruminant Research 104:10–16.

IFAD [International Fund for Agricultural Development]. Programme on Improving Livelihoods of Small Farmers and Rural Women through Value-Added Processing and Export of Cashmere, Wool and Mohair. http://temp.icarda.org/cac/fiber/default.asp

Adventure Yarns. 2011. About Adventure Yarns. www.adventureyarns.com

Community-based beekeeping for better livelihoods

(1) Bradbear N, Joshi DR. 2012. Evaluation report of the project "Improving livelihoods through knowledge partnerships and value chains of bee products and services" [unpublished report]. Thimphu, Bhutan.

(2) ICIMOD [International Centre for Integrated Mountain Development]. 2012. Improving livelihoods through knowledge partnerships and value chains of bee products and services [unpublished report]. Thimphu, Bhutan: ICIMOD.

(3) Partap U, Gurung MB. 2012. Improving livelihoods through community-based beekeeping in Nepal. New Agriculturist. www.new-ag.info/en/research/innovationItem.php?a=2761

(4) Partap U, Partap T, Sharma HK, Phartyal P, Marma A, Tamang NB, Ken T, Munawar MS. 2012. Value of insect pollinators to Himalayan agricultural economies. Kathmandu, Nepal: ICIMOD [International Centre for Integrated Mountain Development].

6 Diversification of mountain livelihoods

(1) Huddlestone B, Ataman E, de Salvo P, Zanetti M, Bloise M, Bel J, Franceschini G, Fè d'Ostiani L. 2003. Towards a GIS-based analysis of mountain environments and populations. Environment and Natural Resources Working Paper No 10. Rome, Italy: FAO [Food and Agriculture Organization].

(2) Nazneen Kanji N, Sherbut G, Fararoon R, Hatcher J. 2012. Improving quality of life in remote mountain communities: Looking beyond market-led approaches in Badakshan Province, Afghanistan. Mountain Research and Development 32(3):353–363.

(3) Jacobi J, Bottazzi P, Schneider M, Huber S, Weidmann S, Rist S [in preparation]. Social-ecological resilience in organic and non-organic cocoa farming systems in Bolivia.

(4) Huber S, Weidmann S. 2012. Das Potential der biologisch zertifizierten Produktion von Kakao zur Verbesserung der Lebensgrundlage der Kakaoproduzenten in der Region Alto Beni, Bolivien [MSc thesis]. Bern, Switzerland: Institute of Geography, University of Bern.

(5) Castella JC, Lestrelin G, Hett C, Bourgoin J, Fitriana YR, Heinimann A, Pfund JL. 2013. Effects of landscape segregation on livelihood vulnerability: Moving from extensive shifting cultivation to rotational agriculture and natural forests in northern Laos. Human Ecology 41(1):63–76.

(6) UNEP [United Nations Environment Programme], Conservation International, Tour Operators' Initiative. 2007. Tourism and Mountains: A practical guide to managing the social and environmental impacts of Mountain Tours. Nairobi, Kenya: UNEP.

Diversification – a historical perspective

(1) Société d'histoire du Valais romand. 2002. Histoire du Valais, Annales valaisannes 2000–2001, Sion, Switzerland: Société d'histoire du Valais romand.

(2) De Riedmatten L. 2004. Le soldat valaisan au service de l'Empereur Napoléon: un service étranger différent (1806–1811). Vallesia 59:1–196.

Small forest-based enterprises reconcile conservation and development

(1) Uganda Wildlife Authority. 2011. Bwindi census of mountain gorillas. Kampala, Uganda.

(2) Plumptre AJ, Kayitare A, Rainer H, Gray M, Munanura I, Barakabuye N, Asuma S, Sivha M, Namara A. 2004. The socio-economic status of people living near protected areas in the Central Albertine Rift. Albertine Rift Technical Reports, Vol. 1. Uganda: IGCP [International Gorilla Conservation Programme], WCS [Wildlife Conservation Society], CARE.

(3) Boffa JM, Turyomurugyendo L, Barnekow-Lillesø JP, Kind R. 2005. Enhancing farm tree diversity as a means of conserving landscape-based biodiversity. Mountain Research and Development 25(3):212–217.

(4) FAO [Food and Agriculture Organization]. 2005. Community-based commercial enterprise development for the conservation of biodiversity in Bwindi World Heritage Site, Uganda. FONP [Forest Policy and Institutions Service], Forestry Department. Rome, Italy: FAO.

(5) Uganda Bureau of Statistics. 2006. Uganda National Household Survey – Socio-Economic Module 2005/2006. Kampala, Uganda.

(6) FAO [Food and Agriculture Organization]. 2006. Community-based tourism: Income generation and conservation of biodiversity in Bwindi World Heritage Site. The Buhoma village walk case study, Uganda. Working Paper No 12. Rome, Italy: FAO.

(7) UN DESA [United Nations Department of Economic and Social Affairs]. Buhoma village, Uganda: Creating new trails in ecotourism. In: Innovation for Sustainable Development: Local Case Studies from Africa. New York, USA: UN DESA.

Social agriculture as part of green care

(1) Haubenhofer D. 2010. Defining the concept of green care. In: Sempik J, Hine R, Wilcox D (eds). Green Care: A conceptual framework. Loughborough, United Kingdom: Centre for Child and Family Research, Loughborough University, pp. 27–35.

(2) Alimos, Alimenta la salute. 2012. Le fattorie didattiche in Italia. Censimento delle fattorie didattiche accreditate. www.fattoriedidattiche.net/images/stories/pdf/fdxreg.nov2012.pdf

(3) Hine R. 2008. Care farming: Bringing together agriculture and health. ECOS 29:42–51.

(4) Sempik J, Hine R, Wilcox D. 2010. Green Care: A conceptual framework. Report of the Working Group on the Health Benefits of Green Care, COST Action 866. Loughborough, United Kingdom: Loughborough University.

Dessein J, Bock BB (eds). 2010. The Economics of Green Care in Agriculture. COST Action 866, Green Care in Agriculture. Loughborough, United Kingdom: Loughborough University.

Di Iacovo F, O'Connor D (eds). 2009. Supporting policies for social farming in Europe. Progressing Multifunctionality in Responsive Rural Areas. Florence, Italy: ARSIA [Agenzia Regionale per lo Sviluppo e l'Innovazione nel settore Agricolo-forestale].

Autonome Provinz Bozen Südtirol. 2012. Qualitätscharta für die Ausübung der „Schule am Bauernhof"-Tätigkeit. PG – act No 526. Beschluss der Landesregierung Nr. 526, Sitzung vom 10.04.2012, und Landesgesetz Nr. 7 vom 19.09.2008. http://lexbrowser.provinz.bz.it/doc/de/6167/beschluss_vom_9_dezember_2008_nr_4617.aspx?view=1

Rural tourism promotion builds on local values

Carpathian Sheep Transhumance Project. www.redykkarpacki.pl/index.php?menu=transhumance&id=&tyt=&j=ENG

Birds of the Carpathians Project. www.ptakikarpat.pl/en/project.html

GotoCarpathia Certification. http://gotocarpathia.pl

Byszewska-Dawidek M, Jagusiewicz A. 2010. Turystyka wiejska w 2010 roku. Instytut Turystyki. Warsaw, Poland.

Szpara K. 2011. Agroturystyka w karpatach polskich [Agritourism in the Polish Carpathians, summary in English]. Krakow, Poland. Prace Geograficzne 125:161–178.

Kurek W. 2004. Turystyka na obszarach górskich Europy, Wybrane zagadnienia. Krakow, Poland:

Instytut Geografii i Gospodarki Przestrzennej UJ.